Urban Planning for Disaster Recovery

Urban Planning for Disaster Recovery

Edited by

Alan March
The University of Melbourne, Australia

Maria Kornakova
The University of Melbourne, Australia
Massey University, New Zealand

ELSEVIER

Butterworth-Heinemann
An imprint of Elsevier

Butterworth-Heinemann is an imprint of Elsevier
The Boulevard, Langford Lane, Kidlington, Oxford OX5 1GB, United Kingdom
50 Hampshire Street, 5th Floor, Cambridge, MA 02139, United States

Notices
Knowledge and best practice in this field are constantly changing. As new research and experience broaden our understanding, changes in research methods, professional practices, or medical treatment may become necessary.

Practitioners and researchers must always rely on their own experience and knowledge in evaluating and using any information, methods, compounds, or experiments described herein. In using such information or methods they should be mindful of their own safety and the safety of others, including parties for whom they have a professional responsibility.

To the fullest extent of the law, neither the Publisher nor the authors, contributors, or editors, assume any liability for any injury and/or damage to persons or property as a matter of products liability, negligence or otherwise, or from any use or operation of any methods, products, instructions, or ideas contained in the material herein.

Library of Congress Cataloging-in-Publication Data
A catalog record for this book is available from the Library of Congress

British Library Cataloguing in Publication Data
A catalogue record for this book is available from the British Library

ISBN: 978-0-12-804276-2

Publisher: Candice Janco
Acquisition Editor: Candice Janco
Editorial Project Manager: Hilary Carr
Production Project Manager: Punithavathy Govindaradjane
Designer: Mark Rogers

Contents

List of Contributors

Iftekhar Ahmed
University of Newcastle, Callaghan, NSW, Australia

Thomas Bamforth
International Federation of Red Cross/Red Crescent Societies (IFRC), Melbourne, VIC, Australia

Edward J. Blakely
University of Sydney, Sydney, NSW, Australia

Esther Charlesworth
RMIT University, Melbourne, VIC, Australia

Kate Cotter
Bushfire Building Council of Australia, Melbourne, VIC, Australia

Catherine Elliott
The University of Melbourne, Melbourne, VIC, Australia

John Handmer
Royal Melbourne Institute of Technology, Melbourne, VIC, Australia

Mark Kammerbauer
Nuremberg Institute of Technology, Nuremberg, Germany

Maria Kornakova
The University of Melbourne, Melbourne, VIC, Australia; Massey University, Palmerston North, New Zealand

Jorge Leon
Universidad Técnica Federico Santa María, Valparaíso, Chile

Alan March
The University of Melbourne, Melbourne, VIC, Australia

Brendon McNiven
Arup International, Melbourne, VIC, Australia

Sarah-Alice Miles
Amersfoort, Netherlands

David O'Brien
The University of Melbourne, Melbourne, VIC, Australia

Janet Stanley
The University of Melbourne, Melbourne, VIC, Australia; National Centre for Research in Bushfire and Arson, Melbourne, VIC, Australia

Mojgan Taheri Tafti
The University of Tehran, Tehran, Iran

Christine Wamsler
Lund University Centre for Sustainability Studies (LUCSUS), Lund, Sweden

Preface

Growing numbers of natural disasters associated with anthropogenic climate change and urbanization have resulted in increased impacts on human settlements over time. Add to this the many other issues faced by dynamically changing urban places and rural areas, and the stresses of disaster impacts become complex and multifaceted. Scholars and professionals in the field now acknowledge the core role that built environment disciplines can play in disaster risk reduction. However, there is often a lack of clear understandings of mechanisms for practical action, combined with shortcomings in understanding the challenges associated with the dynamics of human settlements. Further, there is a need for better understandings of the characteristics of hazards and risks themselves in the built environment.

In this book we take as a starting position that the various dynamics of risk profiles in human settlements are the product of interactions between social, environmental, and built systems, and they need to be addressed coherently in postdisaster recovery processes. However, disaster recovery may focus excessively on single aspects of human settlement systems, such as housing. While such individual elements are clearly important, the treatment of settlements as integrated and multifaceted systems is fundamental to effective recovery, such as the need to maintain and enhance the local economy and social connectedness of citizens. This book partially addresses this gap by providing a theoretical background directly linked to practical examples of successful and challenging recovery processes in the field. Further, we acknowledge the need to see recovery processes as a fundamental starting point and opportunity in preparing for ongoing threats in the future.

The book does not cover all concepts and challenges of recovery for disasters, and we do not propose universal solutions for successful recovery practices. Rather, our goal here is to deal with some of the challenges faced by professionals and communities in the field, to explore role of urban planning and design, and to identify some key ideas and directions for moving forward in approaches to recovery. We allowed the chapters to deal with complex matters, while avoiding highly technical writing to ensure its suitability for a wider audience. The materials covered will be valuable for students, professionals in disaster risk reduction and the built environment disciplines, and the many other citizens and occupants of urban settlements.

This book is a product of collaboration of leading authors and professionals in the field. The idea began with the editors, Dr. March and Dr. Kornakova, along with Dr. Leon—the coauthor of Chapter 1—discussing the need for a text dealing with urban professionals' perspectives of disasters at the 2014 International Disaster and Risk Conference (IDRC) conference in Davos, Switzerland. There we were fortunate to meet a representative of Elsevier, Sara Scott, who was receptive to our approach. The book continued to develop and we would like to strongly acknowledge the role and participation of Dr. Leon in the development of the initial ideas for the book.

CHAPTER

1

Integration and Collective Action: Studies of Urban Planning and Recovery After Disasters

Alan March[1], Maria Kornakova[1,2], Jorge Leon[3]

[1]The University of Melbourne, Melbourne, VIC, Australia; [2]Massey University, Palmerston North, New Zealand; [3]Universidad Técnica Federico Santa María, Valparaíso, Chile

INTRODUCTION

Many of the activities that fall under the broad description of "planning" have characteristics that might be described as a form of governance or collective action—a role that urban planning can at least partly play—particularly insofar as it relates to the management of land, physical systems, and distributions of human activities. This theme is followed in this book, where the focus is upon the suite of actions that represent urban planning, sometimes known as town or city planning. The book examines disaster risk reduction and, in particular, the recovery stage of what is known as the disaster cycle. As discussed later in this chapter, this "cycle," an imperfect but descriptively useful term, acknowledges that disasters occur at given times, but that human actions relating to disaster risk reduction actually occur in one or more of four phases: planning, preparation, response, or recovery.

The theory base used in this book derives from a number of sources in the literature of urban planning, disaster management, and various other approaches to integration and interdisciplinary actions. In parallel, however, it demonstrates that a sustained, evidence-based, and integrated long-term action is fundamental to successful planning outcomes, and that poorly coordinated and planned recovery can lead to long-term risks that could have been avoided. The next section goes on to set out a description of disasters, the characteristics of the recovery, and the ways that urban planning can contribute to recovery.

COPING WITH DISASTERS

Hazards can be defined as natural or manmade conditions that have a potential for social, infrastructural, or environmental damage (Coppola, 2011; Oliver-Smith, 2002). Under certain circumstances, the interactions of hazards with human systems

Urban Planning for Disaster Recovery. http://dx.doi.org/10.1016/B978-0-12-804276-2.00001-3

might lead to disasters, i.e., disruptions of normal functioning of a community or society that involve large human, material, economic, or environmental losses and exceed abilities to cope with the affected society (Emergency Management Australia, 2004; Twigg, 2004; UNISDR, 2009). Historically considered as "Acts of God" or inherent natural phenomena, disasters are now understood as an outcome of social development patterns with higher risks of exposing vulnerable populations to hazards (Mileti, 1999; Oliver-Smith, 2002; Wijkman & Timberlake, 1984; Wisner, Blaikie, Cannon, & Davis, 2004). Consequently, since the 1950s tactics for coping with disasters have steadily evolved from civil defense–based response and relief approaches, with their roots in the "Civil Defense Era" following World War II (Coppola, 2011; Quarantelli, 2000), to risk reduction strategies (Pearce, 2003; Tarrant, 2006; UNISDR, 2004).

At present, general terms, such as "disaster management" or "disaster risk reduction" are used to define standard and organized efforts for reducing harm to life, property, and environment due to disasters (Coppola, 2011). These efforts can increase a community's resilience. Resilience is understood here as an ability to cope with various catastrophes by surviving them, minimizing their impacts, and recovery with minor social disruption (Cutter et al., 2008). In addition, because the type of resilience that planning deals with relates to urban and the "built" environment, it is also useful to consider resilience as stated by Meerow, Newell, and Stults (2016, p.39):

> *the ability of an urban system – and all its constituent socio-ecological and socio-technical networks across temporal and spatial scales – to maintain or rapidly return to desired functions in the face of a disturbance, to adapt to change, and to quickly transform systems that limit current or future adaptive capacity.*

Disaster management processes are usually described according to the components of a cycle comprised four interrelated groups of activities, two of which occur before the catastrophic event and two afterward (Alexander, 2002; Coppola, 2011; Emergency Management Australia, 2004; Topping, 2011; Twigg, 2004):

- **Mitigation** or **prevention**: includes long-term actions aimed at reducing impacts or to eliminate the likelihood of future disasters. These actions can be either structural (e.g., civil engineered defenses, physical retrofitting) or non-structural (e.g., land use planning, insurance, tax incentives, legislation, knowledge development, education).
- **Preparation**: comprises specific measures taken to reduce the impacts of an imminent disaster. Among these are emergency services training, establishment or strengthening of forecasting and warning systems, stockpiling of food supplies, evacuation planning, acquisition of emergency response equipment, preparation of shelters, etc.
- **Response**: consists of emergency actions taken during the impact or immediately after disaster strikes to reduce or eliminate its consequences. They include rescue and relief activities, evacuating and sheltering affected people,

enforcement of public order, resumption of critical infrastructure, coordination of external aid, etc.
- **Recovery**: includes repairing damages and restoring essential community services, restoring community back to predisaster conditions, and creating new opportunities for future development. Among these actions there are re-establishing housing, transportation, public services and economic activity, cleaning of debris, and social rehabilitation programs.

While preparation, response, and recovery are generally sequential activities, mitigation can be conducted at any time (regardless of the occurrence of actual emergencies) (Topping, 2011; Victoria Department of Justice, 2013). The length of response and recovery activities, in turn, will vary from case to case. As discussed further, in large-scale disasters, for instance, the latter may take up to several decades (Alexander, 2002).

RECOVERY AS A "PHASE"

Actions taken during the prevention, preparedness, and response stages can significantly avoid or reduce impacts (environmental, structural, economic, or social) when disaster strikes. The process of rebuilding, repairing, or reconstructing and returning a system to a functional state is referred to as the "recovery" stage of the disaster cycle (e.g., Blaikie, Cannon, Davis, & Wisner, 1994; Coppola, 2011). Aside from rebuilding, the recovery stage aims to restore community to less vulnerable state (Alexander, 1999). Following the response stage, recovery completes the disaster cycle and often shifts back to the prevention stage, acknowledging that some recovery activities are planned for at earlier stages of the cycle (Alexander, 2002). Typically, well-conducted activities undertaken during the recovery stage are the result, in a comprehensive disaster management system, of many activities having been planned in advance, including the allocation of responsibilities and development of regulatory frameworks. Realistically, however, unforeseen situations occur in almost every large disaster event, because of their inherent nature, resulting in unplanned recovery activities. However, if at least some recovery activities are undertaken in planning and prevention stages, time frames will be shortened, particularly relating to the restoration of vital services and facilities.

Recovery follows on from the emergency or response stage, typically only hours after a disaster. However, recovery commonly continues for many years, depending on the level of damage and the resources and capabilities available to a community (Alexander, 2002; Blaikie et al., 1994; Burby, 1998; Clary, 1985). For example, damages resulting from Hurricane Katrina, 2005 in New Orleans, LA, resulted in impacts that have required recovery to continue for almost a decade after the event at the time of writing (Federal Emergency Management Agency, 2012). The prolonged duration of this stage results in certain planning challenges. The early stages of recovery usually involve developing plans for temporary housing, allocating funds for various aids, securing various sources of technical and medical equipment, etc. An important challenge often emerges at this stage, between taking speedy action to resolve immediate issues and achieving long-term disaster risk reduction goals. For example, provision of crisis housing and shelter is typically required within hours or days and consumes significant allocations of land, funds, professional attention, and associated services such as sewerage, food, health care, and so forth. However, as increasing land and resources are allocated to ostensibly temporary, short-term actions, they often impede long-term goals, such as ensuring that the overall layout of a town or city will be more resilient as a result of using the recovery process to redesign key aspects, perhaps including relocation of permanent housing or businesses and associated infrastructure. Awareness of these potential events and activities, therefore, should be included in earlier stages of the disaster management and planning, ideally before disasters occur. These would typically include matters such as development of plans for temporary housing, allocation of funds, and distribution processes and responsibilities for aid, ensuring availability of technical equipment. Of course, the nature of disasters and unforeseen events almost always results in unplanned activities within this stage.

Recovery activities can be divided into short- and long-term actions. Examples of short-term activities are the restoration of vital support systems, providing immediate aid to victims such as temporary housing, medical and humanitarian aid, various financial relief schemes, and recovery of basic facilities and services, such as water and sewerage systems. Examples of long-term activities include rebuilding and reconstruction of the community, thereby aiming to restore physical and spatial images of the community. Reimbursement schemes for property losses, such as insurance, financial aid, or buyback schemes, as well as establishment of ongoing communications with the public and various social rehabilitation programs, are also included in long-term recovery activities (Alexander, 2002; Coppola, 2011).

Recovery is sometimes considered to be the most challenging of the four stages of the disaster cycle. One reason is the need for close collaboration among professionals, agencies, and interest groups from a diverse range of disciplines and perspectives, for recovery to be effective. Recovery must not only include the physical restoration of damaged structures but also initially provide for the safety and well-being of victims in terms of potential vandalism postdisaster, the allocation of funds for restoration works across a range of agencies and people with differing impacts and tenure or other circumstances, surveys of overall damage, clearance and removal works, development of plans for the future, dealing with political processes, and reassessment of hazard risk with new data. Accordingly, it is not uncommon for some actions to conflict with and contradict other activities in the recovery process and, indeed, with other separate urban planning goals (Alexander, 2002; Coppola, 2011). For example, an area might be considered hazardous and highly risky by certain professionals after disaster surveys, but be retained as a residential zone according to regulations that also take into account growth pressures, affordability, and political necessity. These types of conflicts might be resolved by additional assessment, integration between agencies, reallocation of decision-making powers, and development of improved procedures and dispute resolution mechanisms. All these changes fall under the broad banner of governance, of which urban planning plays a significant part. Moreover, governance needs to provide a room for adaptation, acknowledging that while "ideal" recovery processes would be based on plans formulated *before* disaster strikes, in most cases political appetite and resources do not remain strong and sufficient enough to support predisaster reconstruction planning (Alexander, 2002).

Additional recovery challenges exist under the broad banner of "time," particularly the long periods required to develop and implement plans and if significant changes to predisaster states are sought. While recovery plans seek to bring a community back to a less vulnerable state in a timely manner, they typically require several stages of analysis, development, verification, and ratification, including the public. This, however, is often neglected due to time constraints posed by the nature of the recovery stage, coupled with a desire to bring communities quickly back to a functioning state (Alexander, 2002). In most cases, recovery processes tend to rebuild communities in very similar ways as predisaster states (Glavovic, 2010). This does not always allow sufficient changes to take place, again due to time constraints and the need to understand and work with social and community factors. While many residents and recovery managers will generally desire getting back to some sort of predisaster state as soon as possible, measures including ongoing community education and communications need to be established to explain, improve, and integrate changes, to ensure any community is returned to a less vulnerable state.

Another important temporal aspect of recovery is that it typically includes, even if only briefly, what is often seen as the "window of opportunity"—a period during which there is potential to change and improve disaster prevention measures significantly (e.g., Alexander, 1999; Mileti, 1999). Usually a rather short time period, it does not provide many opportunities for change, such as in urban planning, which is an important element of rebuilding processes. It can however, provide opportunities for important regulatory, financial, or cultural modifications to the mechanisms and agencies that underpin urban planning and management processes. The limited time frames for significant changes appear to be due to the political and bureaucratic processes, which contrast with the technical and scientific facts of planning, but are nonetheless inherent to planning powers and implementation. Another reason is the unwillingness of many community members to radically change approaches immediately after an event, when there may be a strong desire to return to predisaster conditions. Despite relief and recovery attempts, "...many people rebuild in precisely the same places and in the same manner so that they remain exposed to recurring events" (Glavovic, 2010, para. 6).

PLANNING FOR RECOVERY

As described briefly earlier, disaster recovery is a prolonged and demanding process that might take years or decades, rather than months. The processes of restoring functionality in a community and of reducing its vulnerability to potential future disasters require collaboration between various agencies and disciplines, adding considerable complexity to activities during this stage. While the range of disciplines is deserving of a separate analysis and a study, this book focuses on one major aspect of recovery—urban planning and related exercises. There is an urgent need to address urban planning and recovery, particularly in a time of rapid change and urbanization. It has been calculated that in the year 2050 approximately 68% (around 6250 million people) of the global population will be urban. It is also predicted that by 2025 there will be 37 megacities with more than 10,000,000 inhabitants in the world, 22 of which will be located in developing countries (Heilig, 2012). In these countries, economic and environmental pressures are driving people from country areas to rapidly forming cities and urban agglomerations, seeking improved livelihoods, leading to rapid and uncontrolled urban growth (Twigg, 2004). Factors, such as population density, poverty, rapid expansion of informal settlements, overcrowding of tenement districts, failure to ensure minimum safety features, and a lack of governance mechanisms (Pelling, 2003), are turning cities into "hot spots" for disasters (Joerin & Shaw, 2010; Wamsler, 2014). Additionally, this problem is exacerbated by exposure of urban populations to biological, chemical, and physical hazards, which, although also existing in many rural locations, are particularly intensified in densely populated cities (Twigg, 2004). For instance, the urban fabric materials and heat emissions aggravate climate effects (e.g., "heat island effect"); unregulated construction, in turn, leads to the occupation of hazard-prone areas, unsafe buildings, and deforestation (Twigg, 2004; Wamsler, 2014).

The strong tendency for disasters to increasingly occur in urban areas must be contrasted with the fact that this is not inevitable, *if* careful planning and management can be introduced and maintained over time. Urban planning for recovery is a core area for improvement that will yield significant returns over time, particularly as it relates to urban areas and their hinterlands. The following section introduces essential urban features that relate to disaster risk reduction practices. Classified in four groups, they provide arguments for further attention and discussion in the various chapters of this book.

URBAN FEATURES AND RISK REDUCTION

PHYSICAL ASPECTS

Physically, urban environments can be managed in three different ways to prevent hazards turning into disasters. *Hazard mitigation* can be achieved through either (1) an appropriate management of environmental or physical conditions that compound/amplify hazards (e.g., vegetation clearance to prevent forest fires) or (2) by building physical defenses to reduce negative impacts of hazards (e.g., seawalls for tsunami protection). The appropriate *location* of urban features separating them from hazardous areas is another preventive mechanism. This can be achieved via (1) restrictions on constructions in vulnerable zones or (2) relocation of existing activities (e.g., moving buildings to higher ground to cope with increasing sea levels). Finally, physical *adaptation* of urban elements has potential to reduce impacts of hazards (e.g., building codes, design guidelines, or subdivision limits for vulnerable zones).

SOCIAL ASPECTS

In a broad sense, the most important ongoing social process influencing disaster risk reduction in cities is rapid urban expansion, which results from natural population growth and rural–urban or international migration (especially in developing countries) (Pelling, 2012; Wisner et al., 2004). This process is usually characterized by high urban densities and rise of social inequalities (Wamsler, 2014) and leads to increased levels of urban vulnerability, especially among those marginalized by gender, age (e.g., young and the elderly), ethnicity, religion, or disability (Twigg, 2004). Poor and marginalized urban dwellers have diminished capacities for coping with disasters, due to factors such as lower levels of education, insecure tenures, lack of social networks, and deficient access to governmental and financial support (Peacock & Prater, 2012; Wamsler, 2014). Additionally, the socioeconomic status and demographic features of a population, combined with levels of experience and education regarding disaster risks, will have a significant impact on the risk levels of a community and the resilience of community members. It is also important to note that this can change over time, for example as a community's demographic tends toward an aging population, or to seasonally high levels of tourists, or as part-time hobby farmers displace more traditional long-term farming communities.

ECONOMIC ASPECTS

Norris, Stevens, Pfefferbaum, Wyche, and Pfefferbaum (2008) argue that a community's economic resilience to disasters is determined by three factors: economic growth, stability of livelihoods, and equitable distribution of income and assets within populations. Twigg (2004) and Wamsler (2014) point out that economic urban risk reduction can be fostered in three different ways. First, a *diversification* and *sustainability* of activities and livelihoods (with proper income levels) has a means to decrease disaster risks. For instance, Adger (2000, p. 354) points out that for a community "dependency on a narrow range of natural resources can increase the variance of income and hence decreases its stability." Second, by ensuring the adequate protection of critical assets and infrastructure [e.g., those "vital to both disaster response and to the overall safety and security of the affected population" (Coppola, 2011, p. 338)]. Third, proper *financial mechanisms* such as insurance and access to grants and credit for implementing mitigation measures have the potential to decrease risks associated with disasters. Disasters can drastically influence the economic situation not only of the affected area, but larger economic communities, such as a state or even a whole country or region, as it impacts not only houses and structures but also the economic assets of the area, such as manufacturing.

ENVIRONMENTAL ASPECTS

UNEP (2005) underlines that "healthy ecosystems often provide natural defences" to hazards (UNEP, 2005, p. 8) and therefore "degraded ecosystems reduce community resilience" (UNEP, 2005, p. 11). Twigg (2004) and UNISDR (2009) also suggest that environmental degradation (e.g., poor management of natural resources and destruction of ecosystems, such as deforestation or pollution) has the potential to increase the frequency and intensity of natural hazards, thereby increasing the overall vulnerability of community. For example, pollution of marshlands and their development in the Mexican Gulf area resulted in decreased numbers of natural hurricane defenses and influenced drastic damage brought by Hurricane Katrina to New Orleans (The Department of Homeland Security, 2006).

Wamsler (2014) argues that interaction of cities with their natural environment might lead to increased disaster risk in six ways: influencing urban climate, creating new hazards via built elements, expanding into hazard-prone areas, compounding new hazards by bringing together a range of competing land uses, producing high emissions, and changing hazard patterns due to dynamic urbanization. On the other hand, adequate environmental management provides opportunity for disaster risk reduction, according to UNEP (2005). It underlines the following: (1) inclusion of environmental change as a parameter of risk; (2) complementation of environmental scientific knowledge with locally based one; (3) protection and value of ecosystem services; (4) combination of engineered defenses with environmental technologies for disaster risk reduction; and (5) strengthening of capacities for environmental recovery. Twigg (2004, p. 249) underlines that "environmental protection or renewal is technically feasible" through activities, such as reforestation, waste management, and sustainable farming and grazing practices, supported by environmental education.

URBAN PLANNING

A brief summary of urban planning is provided here in preparation for a fuller explanation in Chapter 2. Recovery activities after a disaster event must deal with the diverse and complex functions of human settlements, sequencing activities over time in such a way that ensure both short-term and long-term goals of communities are effectively achieved (Australian Emergency Management Institute, 2011, p. 3). As a key part of this, urban planning is primarily concerned with managing the spatial arrangements of cities and towns, in keeping with its ongoing role as urban manager (Hall & Tewdwr-Jones, 2011). This includes, but is not restricted to, the range of physical facilities and structures that settlements need, such as water and sewerage infrastructure, hospitals, schools, offices, industry, and housing. The distribution of these different physical aspects is fundamental not only to the recovery but also to the ongoing functioning of an urban area during "normal" operations. For example, the distance and mode of transport between homes and typical daily destinations, such as work, school, childcare, health, and recreation facilities, is an enduring challenge for urban planning. Similarly, the location, density, height, and arrangements of structures play a key role in determining the ways that an urban place will function. In addition to determining where structures are to be, the interwoven question is to ensure that certain areas are left free of development. This could be to allow sufficient space for food production, aesthetics, recreation, protection of natural areas, or to maintain space for future growth and change. Successful recovery planning needs to integrate and balance the needs and challenges associated with "everyday" and "emergency" operations.

In parallel with the physical aspects of settlements, and implicit in the brief description above, there are many and complex *human* aspects of settlements necessary to their success (Healey, 1997; Keeble, 1952; McLoughlin, 1969). In planning terms, this could be understood in a fundamental way as "land use"—the activities that occur on land and in buildings themselves—and the many links between these land uses. These aspects of use and activity are closely tied with, but separate to the nature and qualities of structures and physical spaces (Bracken, 2014; Lynch, 1984; Tugwell, 1975/1948; Unwin, 1909). Humans have complex needs that include aspects throughout their entire life cycle, from birth to old age, across a range of capabilities and choices. Cultural, recreational, familial, civic, political, and personal aspects of human life need to be accommodated within urban areas and within wider regions. These diverse activities occur within the structures and across private and public domains of urban places and impact upon vulnerability levels and types.

Human settlements are dynamic and always going through changes, whether it be growth and decline, in terms of demographics, economic qualities and fortunes, political and citizen sentiments, quality and life cycle of building stock, to name a few. Accordingly, planning is focused on balancing out and managing current circumstance with future goals and possibilities (Hopkins, 2001) that often coincide with risk management. Accordingly, planning also seeks to integrate management over urban change processes and the development of land and buildings over time (March, 2012). These controls might be based on zones, regulations, master plans, funding, development corporations, or design review committees. Whatever the planning tool used, the key rationale is that while individual citizens or corporations might own land, the achievement of overarching goals for a community, town, city, or region will often require their activities to be directed or even sometimes restricted so that overall benefits can be achieved. For example, an individual might find it lucrative to commence operating a noisy factory on his or her land in a well-serviced and accessible inner city area. However, if this is an existing residential area, it is common that urban planning regulations will prohibit the factory in that location, since the overarching benefits of maintaining residential safety and amenity outweigh a single individual's property rights. Similarly, in risk reduction terms, it is common to discourage development in floodplains to improve the resilience of settlements and the burden on the community when floods do occur, even while many individuals may wish to build there.

Because planning is oriented to the management of change and seeks certain physical outcomes, it achieves this, conceptually at least, via establishment and use of processes providing legitimacy as part of representative government, rational decision making, and shared understandings. These processes include plan making and plan implementation. A range of models exist, set out in more detail in Chapter 2, which seek to use evidence, inclusive, democratic, moral, and rational to make decisions spanning individual and collective matters. Further, these goals and area of jurisdiction span the geographical range from individual sites and structures, though to entire regions and nations. This can be summarized as planning being evidence-based spatial management of settlements to achieve a range of goals that balance out current and future needs, the need for at least some level of equality of process and outcome, and management of expected risks. However, as set out in Chapter 2, these expansive and far-reaching goals also pose a number of challenges and these have considerable implication for the integration of urban planning with disaster risk reduction.

CONCLUSIONS

This book provides an understanding of urban planning so that those involved in disaster recovery can better understand its potential, as well as its limitations and challenges. In parallel, it provides a starting point for those who already work as to planners understand the nature of disasters and particularly the importance of considered action in the recovery phase of disasters.

The following sections of this book set out key areas for attention, complemented by case studies and examples to illustrate ideas and concepts that can be transferred to other settings. The next two chapters set out fundamental principles across the fields of disaster risk reduction and urban planning as they relate to recovery. Following this is a series of chapters that take up the important theme of governance, acknowledging that urban planning derives its powers and acts through organizations and agencies established by and with government.

As discussed in the final chapter of the book, there are limitations to the power of urban planning in recovery processes, and it is important to acknowledge such limitations. However, experience shows we have many opportunities to improve disaster recovery processes and urban planning has much to offer, particularly in terms of providing an integrated framework and range of spatial organization mechanisms for improved recovery and long-term disaster risk reduction.

REFERENCES

Adger, W. N. (2000). Social and ecological resilience: Are they related? *Progress in Human Geography*, *24*(3), 347–364.

Alexander, D. (1999). *Natural disasters*. London, UK: UCL Press.

Alexander, D. (2002). *Principles of emergency planning and management.* Oxford, England: Oxford University Press.

Australian Emergency Management Institute. (2011). *Community recovery* (3rd ed.). Australia: Commonwealth of Australia.

Blaikie, P. M., Cannon, T., Davis, I., & Wisner, B. (1994). *At risk: Natural hazards, people's vulnerability, and disasters.* London, New York: Routledge.

Bracken, I. (2014). *Urban planning methods: Research and policy analysis.* Hoboken: Taylor and Francis.

Burby, R. J. (1998). In R. J. Burby (Ed.), *Cooperating with nature: Confronting natural hazards with land use planning for sustainable communities.* Washington, D.C.: Joseph Henry Press.

Clary, B. B. (1985). The evolution and structure of natural hazard policies. *Public Administration Review, 45*(Special), 20–28.

Coppola, D. P. (2011). *Introduction to international disaster management* (2nd ed.). Amsterdam: Butterworth-Heinemann.

Cutter, S., Barnes, L., Berry, M., Burton, C., Evans, E., Tate, E., & Webb, J. (2008). A place-based model for understanding community resilience to natural disasters. *Global Environmental Change, 18*, 598–606.

Department of Justice, Victoria. (2013). *Emergency management manual victoria.* Available at http://www.oesc.vic.gov.au/emergencymanual.

Emergency Management Australia. (2004). *Emergency management in Australia. Concepts and principles.* Canberra: Emergency Management Australia.

Federal Emergency Management Agency. (2012). *FEMA dollars continue to support new Orleans' Katrina recovery.* online Federal Emergency Management Agency. Available at http://www.fema.gov/news-release/fema-dollars-continue-support-new-orleans-katrina-recovery.

Glavovic, B. C. (2010). The role of land-use planning in disaster risk reduction: An introduction to perspectives from Australasia. *The Australasian Journal of Disaster and Trauma Studies, 1.* Available at http://www.massey.ac.nz/~trauma/issues/2010-1/editorial.htm.

Hall, P. G., & Tewdwr-Jones, M. (2011). *Urban and regional planning* (5th ed.). Abingdon, Oxon, England, New York: Routledge.

Healey, P. (1997). *Collaborative planning: Shaping places in fragmented societies.* Houndmills: Macmillan Press.

Heilig, G. K. (2012). *World urbanization prospects the 2011 revision.* New York: United Nations, Department of Economic and Social Affairs (DESA), Population Division, Population Estimates and Projections Section.

Hopkins, L. D. (2001). *Urban development: The logic of making plans.* Washington, D.C.: Island Press.

Joerin, J., & Shaw, R. (2010). Climate change adaptation and urban risk management. In R. Shaw, J. Pulhin, & J. Pereira (Eds.), *Climate change adaptation and disaster risk reduction: Issues and challenges (community, environment and disaster risk management)* (Vol. 4) (pp. 195–215). Bingley: Emerald.

Keeble, L. (1952). *Principles and practice of town and country planning.* London: The Estates Gazette.

Lynch, K. (1984). *Good city form.* Cambridge, Mass: MIT Press. c1981.

March, A. (2012). *The democratic plan: Analysis and diagnosis.* Farnham: Ashgate Publishing.

McLoughlin, J. B. (1969). *Urban and regional planning: A systems approach.* Praeger.

Meerow, S., Newell, J. P., & Stults, M. (2016). Defining urban resilience: A review. *Landscape and Urban Planning, 147*, 38–49. http://dx.doi.org/10.1016/j.landurbplan.2015.11.011.

Mileti, D. S. (1999). *Disasters by design: A reassessment of natural hazards in the United States*. Washington, D.C.: Joseph Henry Press.

Norris, F., Stevens, S., Pfefferbaum, B., Wyche, K., & Pfefferbaum, R. (2008). Community resilience as a metaphor, theory, set of capacities, and strategy for disaster readiness. *American Journal of Community Psychology*, *41*(1–2), 127–150.

Oliver-Smith, A. (2002). Theorizing disasters: Nature, power, and culture. In S. Hoffman, & A. Oliver-Smith (Eds.), *Catastrophe & culture* (pp. 23–47). Santa Fe: School of American Research Press.

Peacock, W. G., & Prater, C. (2012). Social protection and disaster. In B. Wisner, J. C. Gaillard, & I. Kelman (Eds.), *The Routledge handbook of hazards and disaster risk reduction* (pp. 687–696). Oxon: Routledge.

Pearce, L. (2003). Disaster management and community planning, and public participation: How to achieve sustainable hazard mitigation. *Natural Hazards*, *28*, 211–228.

Pelling, M. (2003). *The vulnerability of cities: Natural disasters and social resilience*. Sterling, VA: Earthscan Publications.

Pelling, M. (2012). Hazards, risk and urbanisation. In B. Wisner, J. C. Gaillard, & I. Kelman (Eds.), *The Routledge handbook of hazards and disaster risk reduction* (pp. 145–155). Oxon: Routledge.

Quarantelli, E. L. (2000). *Disaster planning, emergency management and civil protection: The historical development of organized efforts to plan for and to respond to disasters*. Newark: University of Delaware.

Tarrant, M. (2006). Risk and emergency management. *The Australian Journal of Emergency Management*, *21*(1), 9–14.

The Department of Homeland Security. (2006). *The federal response to hurricane Katrina: Lessons learned*. Available at http://georgewbush-whitehouse.archives.gov/reports/katrina-lessons-learned/.

Topping, K. (2011). Strengthening resilience through mitigation planning. *Natural Hazards Observer*, *XXXVI*(2).

Tugwell, R. G. (1975/1948). The study of planning as scientific endeavor. In S. M. Padilla (Ed.), *Tugwell's thoughts on planning* (1st ed.). Puerto Rico: University of Puerto Rico Press.

Twigg, J. (2004). *Disaster risk reduction: Mitigation and preparedness in development and emergency planning*. London: Humanitarian Practice Network.

UNEP. (2005). *Environment and disaster risk: Emerging perspectives*. Nairobi: UNISDR.

UNISDR. (2004). *Living with risk. A global review of disaster reduction initiatives*. Geneva, Switzerland: UNISDR.

UNISDR. (2009). *2009 UNISDR terminology on disaster risk reduction*. Geneva, Switzerland: UNISDR.

Unwin, R. (1909). *Town planning in practice; an introduction to the art of designing cities and suburbs*. T.F. Unwin.

Wamsler, C. (2014). *Cities, disaster risk and adaptation*. New York: Routledge.

Wijkman, A., & Timberlake, L. (1984). *Natural disasters. Acts of god or acts of man*. Washington: Earthscan.

Wisner, B., Blaikie, P., Cannon, T., & Davis, I. (2004). *At risk: Natural hazards, people's vulnerability and disasters* (2nd ed.). London, New York: Routledge.

CHAPTER

Urban Planning and Recovery Governance

2

Alan March[1], Maria Kornakova[1,2], John Handmer[3]

[1]The University of Melbourne, Melbourne, VIC, Australia; [2]Massey University, Palmerston North, New Zealand; [3]Royal Melbourne Institute of Technology, Melbourne, VIC, Australia

INTRODUCTION

When disasters occur in and around human settlements, their consequences are highly place-specific, revealing how a particular hazard interacts with and has consequences for the way humans have built upon and live in given locations. Once a disaster has occurred, the process of recovery challenges urban planning, a discipline charged with the management and ongoing improvement of our settlements, by presenting an opportunity to reconsider and improve upon a settlement's characteristics. We present in this chapter a way of thinking about urban planning during the recovery phase that is procedural, as well as being future-oriented to improved disaster resilience as an ongoing process.

A common approach used in recovery is the principle of "building back better." Oriented to a whole-of-community risk reduction approach, it also seeks to reduce the likelihood and consequences of future disasters, as well as rebuilding. It is intuitively logical as the possibility of ongoing losses from repeated events is very real. For example, in the United States the proportion of repeated payments under the National Flood Insurance Program is approximately 25% of the dollar value of payouts, despite being just 1.3% of the total number of policies held (Cleetus, 2015, p. 4).

Establishing planning processes and governance that facilitate improved risk profiles during recovery is now understood as a core goal of disaster risk reduction. The Sendai Framework for Action 2015 review of Build Back Better highlighted the centrality of good governance prior to events to recovery, and that:

> *Recovery needs to be viewed holistically - as part of a continuum, inseparable from preparedness, response, mitigation, and sustainable development. Moreover, recovery must be approached in a cyclical nature wherein actions to strengthen resilience are taken both before and after disasters occur – rather than a linear approach that limits recovery action to the aftermath of an event*
>
> **UNISDR (2015b)**

Despite the logic of reducing future disaster losses and general acceptance of the concept, it remains problematic in practice, particularly within the complexities of urban planning practice and its governance settings that are not generally oriented to

Urban Planning for Disaster Recovery. http://dx.doi.org/10.1016/B978-0-12-804276-2.00002-5

disaster risk reduction (DRR). In particular, there is often limited agreement regarding what "better" means and there are typically multiple views on how to achieve outcomes across various recovery scenarios. Even if consensus is reached, the construction sector might not have the capacity or inclination to deliver radically new designs and materials in quantity—as happened in Australia after the Black Saturday bushfires (Kornakova, 2016). In another case, after a devastating wildfire in 2014 in Valpariso, Chile, nongovernmental organizations (NGOs) rebuilt many houses immediately, recreating the vulnerabilities of the previous settlement. Alternatively, following the 2004 Indian tsunami, many small fishing settlements, often with informal tenure, were relocated away from the sea for safety (e.g., Bavinck et al., 2015; De Silva & Yamao, 2007). However, this disrupted the livelihoods of surviving residents and, in some cases, led to increased risk when the cleared set-back areas were rebuilt with tourist facilities. Furthermore, a focus on buildings alone can overlook financial, social, and economic issues. Local livelihoods and local economic activity are particularly important, such as in Christchurch, New Zealand, in which the post 2011 earthquake central business district cordon displaced about 50,000 jobs.[1]

We suggest that urban planning, understood as a form governance, can play a significant role in improving pre-event conditions and post-event rebuilding that are beneficial to effective recovery. In particular, planning can act as a highly effective spatially realized knowledge and decision base for ongoing improvement and decision making, as well as building governance capacity. As suggested by the UNISDR (2015b, p. 3):

> *[T]he single most effective decision a community or country can make to ensure efficient and effective recovery is to strengthen government systems for recovery before a disaster strikes, through pre-disaster recovery planning. During much of the actual recovery period, many decisions will require split-second action that allows little or no time for analysis. A pre-disaster plan or strategy outlining overarching goals and objectives can help guide post-disaster planning, and reduce the likelihood of ad-hoc behaviors or decisions. It can ensure that pre-existing vulnerabilities are addressed and disaster risks are reduced.*

Urban planning can increase certainty that recovery will improve risk profiles by pre-establishing building standards, overall allocations of services and infrastructure across communities or regions, and the assumption that risks in existing settlements are within acceptable standards. However, this certainty may be at odds with actual practical abilities to modify the strategic direction of growth and change in the recovery process. Further, the prior manner of development remaining after an event (such as street patterns and location of activities, for example) may be at odds with contemporary standards of risk, ecological, social, and economic sustainability. Accordingly, practically achieving ideal standards in terms of risk reduction and sustainable development may have to be traded off with the capacity of local industry and government to deliver. Notwithstanding these challenges, urban planning offers considerable potential for risk reduction and recovery. While it is valuable to appreciate the outcomes of good urban planning

[1] See Chapter 6.

[see Sendai Framework (UNISDR, 2015c)], there is a parallel and ongoing need to understand *what* planning is and *how* it actually occurs. The next section sets out key planning processes and mechanisms, particularly as they apply to developed and democratically-oriented countries or those in transition.

URBAN PLANNING: RECOVERY AS *PROCESS*

A core goal of mature planning systems is to bring about advantageous spatial arrangements of all the physical and functional features in urban and regional areas. These might include housing, recreation, health services, infrastructure, transport, education, industry, and so forth (Halligan & Power, 1992). It is noteworthy that while this understanding tends to focus on physical matters, such as the design and location of structures, the purposes of planning are equally oriented to improving social, economic, and ecological outcomes via these physical processes.

Many models of urban planning processes exist and these continue to be contested and refined over time. Importantly, the most enduring procedural model has its origins in the work of Patrick Geddes, known by the shorthand of Survey-Analysis-Plan (Buxton, Goodman, & March, 2012). This was further developed by various others and applied at multiple spatial scales, for many purposes, notably Mumford (1968) who championed evidence-based strategic planning and action at the regional scale, Abercrombie (1943), and later Lewis Keeble at the metropolitan, town, and precinct scales (Keeble, 1952). The model at the core of these approaches is known as rational comprehensive planning (RCP), although diverse and important variants exist, notably systems planning (McLoughlin, 1969). The model was adapted to urban planning by Meyerson and Banfield (1955), seeking to reduce the negative influences of corruption, subjective values, and politics upon planning processes, in favor of a more scientific and evidence-based approach. The key steps of RCP (as adapted in Keeble, 1952; Taylor, 1998) are:

1. analysis of situation and identification of problems/opportunities;
2. identification of alternative goals and objectives;
3. design of alternatives;
4. comparative evaluation and selection of alternatives against goals;
5. implementation; and
6. monitoring of effects and adjusting goals or other parts of the process.

Even while RCP is periodically criticized, augmented, or ignored since first being developed in the 1950s, it remains relevant (Hoch, 1994; Levy, 2000; Sandercock & Kliger, 1998; Yiftachel, 1999, p. 21). It is noteworthy that the international standard for risk reduction processes *Risk management – Principles and guidelines* (ISO31000, 2009) is an adaptation of the RCP approach. Alternatives such as incrementalism (Lindblom, 1965) are more pragmatically based on the "reality" of managerial approaches that adopt the approach that ongoing adjustment to plans are needed. While perhaps appropriate in certain implementation processes, this approach is criticized for its complexity and attendant loss of strategic and collective oversight, and more fundamentally for its appropriation by various bureaucratic, political, business, or interest groups. While a popular approach as an ideal, the impacts of *realpolitik* often make this ideal implausible, although its use in plan making rather than implementation settings remains more realistic.

As set out in Chapter 1, recovery is generally understood as the process of rebuilding, repairing, or reconstructing and returning a system to a functional state after a major event or shock (e.g., Blaikie, Cannon, Davis, & Wisner, 1994; Coppola, 2011). In parallel with physical rebuilding, this stage is one in which attempts can be made to improve the resilience of communities (Alexander, 1999), including a range of social, economic, and ecological measures. Central to urban planning's ability to add to recovery processes and make improvements to the functions of settlements is an understanding that urban planning must always occur as a *process*, even while some end point ideals of settlement design and function may be a key aspect of directing planning's activities.

While boundaries can become blurred, planning processes broadly occur as either *plan making* or *plan implementation*. Plan making involves all of the prior activities associated with envisaging, testing, and selecting desired future outcomes, for example, the preparation of a policy and associated regulations to restrict and improve construction on sites subject to landslip. This would be based on initial technical and evidence-based analysis, usually with mapping and spatial components. The *implementation* of this policy might include mapping and then applying the regulations to the correct sites and the assessment of individual land owners' applications for development against the criteria of the policy.

The outcomes of implementing planning policy, if successful, might be that future development on highly risky sites would be precluded and that development on moderately dangerous sites would be managed and designed to reduce risks to a manageable level. Development control processes are commonly used to require that the design, materials, and layout of physical structures are appropriate. They also ensure that people occupying these areas are aware of risks and are capable of evacuating in a timely manner or of effectively responding to a disaster. Additionally, sites with low risks would also be identified so that streamlined processes could be established for these areas. For example, an individual might apply for a permit to build a house on a site with moderate landslip risks. The assessment procedures would determine the characteristics of the risks on that particular site, including seismic, engineering, geotechnical, and other expert inputs, and issue a permit subject to conditions. The conditions might include restrictions on excavation and vegetation clearing, engineering requirements for the structure, earthworks and water management on the site, and so forth. The benefits of such a process would be that overarching (including non-hazard–oriented) goals such as managing development pressure and aesthetics would be achieved, while also managing risks to individuals and the community to a level that is quantifiable, economical, and considered acceptable when assessed against a range of tests.

PLANNING USING EVIDENCE

In developed countries, the practices of urban planning use various sources of knowledge and evidence as a base for developing future directions, to provide guidance in recovery processes, with rational reasons providing legitimacy. Among these sources are various data sets, past experience, professional and personal knowledge, interactions with other agencies, decision makers, professionals, and community members to name a few (Krizek, Forysth, & Slotterback, 2009). This places the development, assessment, and application of various evidence as a core element of practice. Moreover, the holistic nature of planning requires a multidimensional approach to practice as it includes a range of different systems, at different spatial scales. Analogous to a living organism, any city consists of many diverse systems, such as transportation, water supply, infrastructure, waste removal, energy provision, housing supply, economic production, health provisions, and many others. These need to be integrated to be sustainable, particularly since considerable amounts of resources are consumed in developing and maintaining a city. Therefore, evidence in planning is core to justifying the multidisciplinary elements underlying the decision-making processes informing planning. For example, informed choices about the release of new land after an event in a particular location will have implications for costs in terms of infrastructure, housing, affordability, distance to places of work, loss of habitat, requirements for new schools, health care, and so forth. Importantly, the nature of urban development may have implications for disaster risk management, meaning that hazards need to be fully understood and risk profiles developed for any proposed and existing settlements. This section briefly introduces evidence in planning and provides several examples with particular focus on hazard mitigation.

Put simply, urban planning is an exercise in urban management based upon spatial understandings such as mapping being combined with various other types of analysis and action (e.g., setting parameters for future growth patterns) being taken on this basis. This requires the gathering and analysis of various data sets to provide a sound base for understanding the benefits and implications of the many possible futures a human settlement might take. These data sets will include topography of the area, population trends, transportation assessments, economic trends, and so forth. Topography is usually presented in spatial data sets, also referred as geospatial data or geographic information, identifying the location of features and boundaries of natural and constructed features. With ongoing development of technologies, these data are becoming more detailed compared with previous survey based data. As a part of mapping exercises, spatial data are analyzed and converted through various software packages and can further be used by planners for development or updating of various maps on different levels.

When applied to DRR and recovery techniques, spatial data can be used to identify vulnerable areas based on geographical features of analyzed areas. For example, growth trends could be identified and combined with mapping that model flooding levels and velocities, leading to development of policies that direct growth to more suitable areas and specify particular building standards. It could also be used to plan evacuation routes while ensuring road capacities are sufficient, to find appropriate locations for future development, or to ensure that sufficient distances are maintained downwind from a potential hazard such as an oil refinery. Spatial data and modeling can determine locations suitable for development, including matters such as further analysis of soils, for example, to allow avoidance of landslip hazards, combined with the use of building and construction codes, and provision of detailed engineering inputs. Chapter 7 provides more detailed practical example of evidence applied planning in Switzerland, demonstrating the need for multiple evidence sources for planning for DRR.

Empirical (used here to mean "observed") data are another example of evidence being used in planning to inform choices about the management of cities and regions over time and, particularly, in the recovery process. For example, statistical analysis of populations allows identification of trends in general or of specific communities (e.g., growth rates, health differences between parts of the city, transport preferences). This also allows identification of the likely demands and needs of a community projected into the future, selection of the most beneficial growth, and change strategies. For example, if a population is aging significantly over time, it may be appropriate to provide different housing types in the future to complement that population's needs, combined with health facilities catering to elderly health care.

Empirical data can be used to influence development strategies, resulting in modified economic strategies, changes in planning and building requirements, land use and zoning of the area, or other actions. Importantly, when applied to urban planning focused on disaster management, empirical data allow professionals to estimate vital elements that can assist in the prevention stage. For example, the number of households likely to require evacuation assistance could be modeled for a proposed new housing development, allowing an informed decision about the risks associated with it to be understood and dealt with or simply avoided in advance. For example, recent changes in wildfire planning in Victoria, Australia, require provision of defendable space and removal of nearby vegetation for new homes, based upon a combination of wider mapping, detailed site assessments, and setting of site-specific building standards based upon likely future risks.[2]

Evidence from different fields and disciplines allow planners to create various scenarios of possible future events and test them out to identify best available solution. This is particularly evident in the creation of evacuation routes and refuge points in advance of future potential disasters. Spatial data and any previous event histories allow professionals to establish potential future disaster characteristics, which can be further added to community maps. As discussed in the next section, however, this typically requires inclusion of many other parties and collaborative approaches with other stakeholders.

---

[2] See Chapter 10 for more details.

GOVERNANCE AND RECOVERY: PLANNING WITH AND FOR OTHERS

Governance in democratic nations is generally understood as the wider set of processes that bring about collective outcomes, including but not restricted to, the formal agencies and institutions of government and often the public and interest groups (Healey, 1997). In this sense, it is also understood as being based upon good process that builds capacity in the wider citizenry in addition to a narrower view of "correct" decisions and outcomes. This wider view acknowledges the value of group learning, the development of trust, and the need to understand and acknowledge the views of diverse stakeholders. The processes of recovery, however, also place particular demands on the need for decisiveness and strong action, meaning that trade-offs need to be made (March, 2012). So while good process is important, it might not always be enough in disaster recovery. Recovery, and disaster management, is typically assessed against outcomes, as well as processes. An argument can be made that this is reasonable and what is expected by all the stakeholders in recovery.

Tierney (2012, p. 344) articulates the connection between governance and impact reduction:

> *disaster governance consists of the interrelated sets of norms, organizational and institutional actors, and practices (spanning pre-disaster, trans-disaster, and post-disaster periods) that are designed to reduce the impacts and losses associated with disasters arising from natural and technological agents and from intentional acts of terrorism.*

We might dispute aspects of this, in particular the conflation of terrorism with disasters inter-related with natural processes, but the overall picture aligns well with understandings of good governance, as set out earlier—with an important exception being the emphasis on outcomes. Similarly, at the global level, the United Nations Office for Disaster Risk Reduction (UNISDR) emphasises outcomes in its 2011 Global Assessment Report (UNISDR, 2015a, p. 116) and noting also that good disaster governance is not easily achieved, concluding that "aside from reducing disaster mortality, existing risk governance capacities and arrangements generally fail to achieve their aims."

Although definitions of governance generally include all formal and informal means of management by government and organizations fulfilling key social and economic roles, most discussion, ranking, and commentary draws on formal institutions, especially the institutions of government. Commerce, civil society, and NGOs are too often absent even though they are important at all stages of DRR and management and are key to recovery governance and outcomes.

TOWARD GOOD DISASTER RECOVERY GOVERNANCE

We acknowledge that urban planning is only one aspect of overall governance processes in recovery. Governance is a complex and dynamic set of mechanisms that extend well beyond the formal agencies of government and political influences.

In the case of recovery, there are the added stresses of time pressures to help those affected, the emotional, and practical strains associated with dealing with the aftermath of a significant event and the potential for finding ways to improve risk profiles during recovery. Urban planning is often a key factor in these processes, with its potential to influence physical and spatial outcomes in ways that impact significantly upon social, economic, and environmental concerns.

Understanding the quality of governance and finding ways to improve it is key to effective recovery. As a starting point, the Overseas Development Institute (ODI) and United Nations Development Programme (UNDP) have developed and applied a disaster risk governance index (Wilkinson, Comba, & Peters, 2014). It starts with the position that human development, political stability, and democracy are needed for good disaster risk governance. However, the resultant index score depends on the details of the combined index construction (Wilkinson et al., 2014, p. 12). The index rates governance by measuring both disaster-specific actions, such as plans, regulation, and policies, and more general attributes, such as accountability, transparency, and participation. The resulting index ranks many countries highly, yet many achieve considerably lower scores, such as Vanuatu in the South–West Pacific region.

The index is based on three existing indicators with global coverage that focus on "generic governance characteristics, and environmental shocks and stresses" such as those from disaster risk management and climate change adaptation. The three indicators comprising the disaster governance index are:

1. coping and adaptive capacities as measured in the World Risk Report (Alliance Development Works, UNU-EHS, & The Nature Conservancy, 2012), including perceived corruption index, good governance (failed states index), various medical facility and health outcome indicators, and a range of capacity indicators such as literacy rates and natural resource management;
2. the readiness score Notre Dame Global Adaptation Index (ND-GAIN), that is, the national level scores of vulnerability and readiness to adapt to climate change, consisting of economic, governance (e.g., accountability, stability) and social indicators (e.g., education, mobile phone usage, rule of law); and
3. the national monitor for the Hyogo Framework for Action (now Sendai Framework)—indicators from all five priority areas are included. These are that DRR is a national priority with capacity for implementation; risks are identified, monitored, and with early warning systems; a culture of safety is developed; risk factors are reduced; and response capacity is strengthened.

The top ranked countries are mainly those with high levels of human development, such as western European, North American, and Australia. However, there are exceptions—disaster recovery outcomes in Italy and Turkey have been poor despite solid performance on the index, which might be a result of a series of very destructive earthquakes. Lower scoring countries tend to have problems implementing disaster risk governance and find that finance and expertise are limiting factors. However, there are exceptions here as well—Cuba has been observed as an exception in development terms as it has long shown high capacity and positive outcomes in the area of disaster risk (Wilkinson et al., 2014, Box 1).

Good governance emphasizes local participation and power. There is a question of how much influence local communities should have on national/state governments, especially those with centralizing tendencies. For example, the US and Japan local governments typically have much more autonomy and responsibility compared with other similar countries such as Australia, and nationally the countries score high on all aspects of the ODI–UNDP index. However, the case of Hurricane Katrina in New Orleans stands out as an example of poor disaster risk governance at every stage including recovery—processes were lacking and the outcomes remain poor. Limited resources and less than favorable formal arrangements do not have to result in poor governance and outcomes.

To summarize the arguments above to the extent that this is possible: high scores on indexes emphasizing democratic participatory processes do not guarantee a sound recovery, and at an individual event level this is far from the case. Recovery governance needs to include a focus on outcomes, and those outcomes need to be achieved for the people involved within a reasonable time frame. This is far from straightforward. There is a real risk that property owners find themselves unable to rebuild or use their land, entangled in arguments about insurance, and generally caught in a context of uncertainty and indecision, which can be exacerbated by open-ended processes and legal arguments. This potentially has negative implications for local livelihoods and economies. As governance is concerned with both processes and agencies, the latter must be explored to establish clear understanding of good governance.

URBAN PLANNING AND THE MAIN AGENCIES IN RECOVERY

A key theme of this book is recognition that urban planning can be a key mechanism oriented to achievement of collective outcomes, focused here upon DRR in the recovery phase. However, it is important to note that "balancing" is often required between individual and group outcomes and control, and between governmental and individual responsibility. In this sense, urban planning processes can sometimes result in reallocations of rights away from individuals, in favor of overall benefits such as to future generations, including risk reduction.

While various sectors have significant roles to play, the main medium through which urban planning achieves collective outcomes is via government agencies. These include local authorities, municipalities, and town councils at the local level that prepare future development plans and maintain and enforce land use regulations or zoning rules. They often carry out important public projects integrating the needs of human settlements, such as building levees, building standards, vegetation management, and development of emergency management systems and warnings. In parallel, there are typically metropolitan-, state-, and national-level agencies with planning responsibilities and influences, such as state or national planning departments. In addition, many stand-alone agencies exist. These are usually oriented to particular tasks, such as housing, roads, redevelopment of key sites, or natural resource management, defense, and security.

A key aspect of western urban planning traditions and practice, as it is facilitated by government agencies in the recovery phase, is its legislative underpinnings. To provide legitimacy for agencies to carry out their roles, it is typical that a legislative base exists to provide a range of powers, responsibilities, and procedures. This legal foundation enables various agencies to act and to make decisions that often cut across the property or personal rights of individuals to build upon or use their land. From this basis many of the main processes of planning, such as the mechanisms for preparing plans and implementing them, have key statutory elements. These will include matters such as the notification of land owners and occupiers if changes to planning regulations are to be made, provision of key information, lodgment of objections and appeals, and final decision procedures. In addition, it is common that there will be a legislative basis for the interactions between the various agencies relating to planning, such as transport, waterways, natural resources, education, and health.

Recovery processes can be contrasted with the relatively "routine" nature of ongoing planning, even while the core goals of providing for human needs in a sustainable way remain, albeit under different circumstances. Recovery processes are typically facilitated by a specially established agency with the far-reaching powers required for the management of the complex tasks associated with this phase of a disaster. Often established by an act of parliament, or via powers already embodied in an emergency or disasters act, these recovery agencies are provided with extensive powers and finances that mirror the complexity and scope of the tasks associated with recovery. Many of a recovery agency's tasks include matters directly or indirectly associated with urban planning. These include location of temporary shelters and reconstruction, reestablishment of utilities and services, measures to reduce ongoing risks, and the development and implementation of overall reconstruction plans involving multiple actors and processes. Table 2.1 shows a simplification of the main actors and agencies by approximate governance tier. It also should be noted that sometimes there are some inconsistencies and problematics of power distribution associated with ad hoc nature of recovery processes. Larger organizations come into the process without clear understanding of context, goals, community needs, etc. and may be overly oriented to their own internal approaches and the expense of wider coordination. In larger disasters there is an ongoing issue of number of such organizations and their uncoordinated inputs, which often influence long-term recovery processes.[3] We argue that good governance and clarity of roles has the potential to address these.

URBAN PLANNING AND RECOVERY: POTENTIALS AND PROBLEMATICS

Within the broad parameters of the agencies and groups outlined above, it is clear that urban planning is but one of many actors. However, we argue here that it offers an important toolkit that is integral to effective recovery. The final section of this

[3] See discussions in Chapters 3 and 11.

Table 2.1 Typical Agencies and Groups in Recovery by Approximate Governance Tier (Indicative Only)

Level	Agencies and Actions
International or multinational	*Agreements* for aid, finance, expertise, and direct assistance. These are often set in place and updated over time as part of wider agreements between nations alongside other matters, such as military support, free trade, work right, and visas, but may occur in an ad hoc manner.
	Nongovernmental organizations. These can take on significant roles that include significant autonomous actions, depending on the nature of the event. It is common for large NGOs to act quite autonomously or to be given significant powers in developing countries where government services have been overwhelmed.
	Religious organizations. These often have strong networks and connections and actively participate in recovery processes, especially in developing countries.
National	*Allocation of disaster relief and reconstruction funding*. This is typically managed and distributed by higher tier national agencies, usually directly tied to reconstruction agencies and activities. While it is typically associated with national and state-level declarations of disaster, it may also be associated with local actions, benevolent funds, or international agencies.
	Legislative basis for national or federal funding and action (e.g., declaration of emergency/disaster).
	Insurance corporations. These play an important role in the recovery process and may form agreements with reconstruction authorities to facilitate equitable risk reduction in recovery.
	Research and science institutions. These play a role in providing a credible evidence base and may play a role in providing ongoing development of knowledge and improvement that can be taken up during recovery processes. (They may also exist at upper and lower tiers.)
	Military. They often provide immediate relief and coordinate recovery; in some countries (e.g., the United States), they also participate in prevention processes (e.g., levees are often under military jurisdiction).
State/provincial/ regional	*Spatial resource allocation agencies* (nondisaster operations) exist at this tier. These may include large-scale infrastructure and planning agencies: urban planning and population distribution, water, sewerage, drainage, energy generation and distribution, roads, public transport, and so forth. During recovery, these agencies often take on extraordinary roles or form special panels that fast-track delivery of tailored planning and other outcomes.
	Social, medical, welfare, and other human-orientated agencies. These, associated with human well-being, are funded and have regulatory powers at this level, such as housing, welfare, and other services.

Continued

Table 2.1 Typical Agencies and Groups in Recovery by Approximate Governance Tier (Indicative Only)—cont'd

Level	Agencies and Actions
	Review boards/boards of enquiry or royal commissions/ inquests into the causes of and reasons for disasters. While these agencies may be legal in their basis, they may have far-reaching powers of recommendation and reform with impacts on recovery. They often override "normal" legal and planning powers. They may integrate and use the knowledge of expert committees.
	Response agencies. These may often continue to have inputs and may take a significant role in recovery, often on the basis of strong legislation and allocated funding.
	Extraordinary recovery agencies with special powers. These are typically headed by ministers, commissioners, chairs, directors, and boards. They play a central role in integrating and regulating activity as a central organizing body with extensive discretionary powers that may override normal processes. They may have a statutory basis in emergency or response agencies.
	Private sector companies. These, contracted to clean up and/ or reconstruct, provide services (e.g., water, telecommunications) or establish funds and programs.
	Local government. This is often the main point of contact and delivery of services to local communities; even if they are overwhelmed by an event, they are often supported by other agencies. It is noteworthy that is common for many planning functions to be delivered or administered by local government, and this extends to recovery processes.
Place or disaster event scale	*Grass roots organizations*. These may form spontaneously or develop from existing voluntary or other preexisting groups. Often formed around shared interests or concerns, these groups may endure and mature in the long term or continue only if the need remains. Examples include religious, educational, sports, social, and other volunteer organizations. It is often the case that grass roots organizations form around planning and related issues in the recovery processes, such as the introduction of modified regulations that impact upon rebuilding.

chapter sets out the main mechanisms of urban planning, integrating these with a brief discussion of the potentials and problems associated with these in recovery. This provides a framing for the more practical examples provided in subsequent chapters.

The process of *vision building* is central to many aspects of urban planning (Hopkins, 2001). If successful, this production of shared purpose, perhaps projected forward 30–50 years into the future, is a powerful way to draw together disparate actions towards core organizing principles. Some of the most enduring city plans and designs have been undertaken on the basis of powerful visions, such as Paris' boulevards, Copenhagen's "Five Fingers," and the Greater London Plan. It is noteworthy, however, that these examples also represent planning and planners being in positions of power and influence that are difficult to reproduce in today's settings. Further, in the immediacy of disaster recovery situations, it is often challenging for people and agencies to be able to project forward beyond more pressing concerns. We suggest that it is usually more appropriate to revisit only the elements of existing visions that might need to be modified to achieve risk reduction goals in the period of opportunity that exists after an event to rebuild in better ways and to allocate funding and greater political will in the most effective way. It should be acknowledged, however, that risk reduction goals can result in the need to change the entire vision of the community to seize opportunities and to maximise learning from past events. In such cases, there is a need for more extensive collaboration with other professionals and community.

Strategic planning is the process of directing or redirecting and integrating elements of ongoing processes, usually on the basis of maintaining and overview of the spatial and functional elements of settlements and the manner in which they are dynamic or static (Hopkins, 2001). This is where the often heard mantra of "getting back to normal" or "bouncing back" may be at odds with the most appropriate (and new) course of action. After a major disaster, returning to normal or to what was there before is often not an option, or is ill-advised due to the need to modify settlement patterns and urban areas so that the initial circumstances that brought about the event are modified. There are circumstances where disasters are far from abnormal, for example, earthquakes in New Zealand or droughts in inland Australia. In other cases "normality" may be a contributing factor to the community's vulnerability to the disaster. Traditionally, psychosocial and environmental recovery aimed for the restoration of "normality" or the predisaster state, or a limited improvement on this. In contrast, infrastructure and economic recovery offer the opportunity for substantial, strategic improvements. For example, destroyed infrastructure and housing are frequently replaced with up-to-date facilities, and local commerce may receive new equipment and training. Often, restoration may not be possible: people may be left with permanent injuries or trauma, parts of the local economy may not be able to reestablish, and the area may be stigmatized as a scene of tragedy. We argue that "normality" can be changed to the "desired state" of the system, as discussed by Meerow, Newell, and Stults (2016), and further move toward the "new normal"[4]. Nevertheless, opportunities for major change and economic enhancement may present themselves, especially where similar disasters are not frequent.

Urban planning processes are often the most powerful in that they represent a mechanism to establish wider decision-making criteria on the basis of *policy and regulation* (Hopkins, 2001) and are often integrated with legitimizing governance structures, such as representative democracy and funded government agencies. Policies are statements of intent, such as "we will not allow new development in areas designated 1:100 year flood zones" and will be backed up by law, regulations in ordinance, and processes of decision making that draw on multiagency

[4] See Chapter 5 for further discussion.

consultation and the application of technical and professional expertise. This may also include opportunities for consultation and citizen inputs. In recovery this may provide significant strength to achieve improved risk profiles, but existing policy and regulation may also have put in place decision-making processes that reinforce the reproduction of risky settlement patterns. Further, many parties may have strong vested interests in maintaining pre-event policy settings to protect financial or other interests. Another potential outcome, as discussed in Chapter 10, is driven by the general desire of the community to get back to "normal," which places additional pressure on decision makers and new regulations, codes, etc. that may be adopted in a rushed manner, often without careful consideration and considered inputs from the science community.

A key tension that emerges across the recovery literature is the challenges between using current regulations and processes, which may no longer be appropriate, allowing sufficient and fundamental autonomy to local communities to ensure capacity is improved and the need for extraordinary powers to be allocated to rebuilding or recovery agencies to give them some chance of being effective. Accordingly, pre-event decision making, agency integration, and citizen participation practices that may be written into regulations may need to be modified significantly. Importantly, it is important here to note that opportunities to improve settlement patterns and to develop new risk profiles will often be tensioned against the need to maintain livelihoods, to return to "normal," and to achieve targets set by government. This may be further complicated by the presence of outside agencies, such as NGOs, and researchers who would not normally be present and who typically seek to achieve specific targets within set time frames (e.g., housing rebuilds and of infrastructure works).

Agenda or project planning is based on planning agencies having a key role in the location, type and timing of major projects, facilities, and infrastructure (Hopkins, 2001). In recovery this might include new flood walls, additional clearing of forest fuels, improved response infrastructure, and so forth. However, the recovery literature is replete with examples of inappropriate and unwanted structures provided after a disaster. These are typically provided according to a standard design from another time and place and take (often intentionally) little notice of local preferences, imperatives, or needs. Apart from sometimes violating cultural norms, imposed housing and relocation often separates people from their livelihoods that are key to their identity and survival. The literature therefore not surprisingly has long supported strong local involvement, if not leadership, in the recovery process for pragmatic, as well as ethical reasons. Nevertheless, the reality in many major disasters around the world in that local involvement is weak regardless of the rhetoric.

The need for good project management in recovery is paramount, balancing various expectations such as speed in reconstruction and restoration versus reducing risk and meeting other agendas (some of which might not be those of local people, such as lengthy exclusion from damaged areas). This also emphasizes the importance of project management in recovery. This may include other factors, such as equity and fairness: both in procedural and representation facets, and in term of outcomes, even while this may be complicated by issues of uneven insurance, government support, and tenure type. Most models of, and commentaries on, recovery emphasize the importance of involving the affected community in its own recovery—ideally in a position of strong influence or leadership. However, despite this some models or aspects of recovery models assume strong leadership or control by the state or external NGOs. As much of the recovery assessment literature relates to developing countries, it often assumes that the government has limited capacity or is replaced in part by NGOs.

Local involvement and control is however not without its problems. For example, in Aceh, Indonesia, where it is reported that people would use key structural components and cement to extend and decorate the dwelling, rather than for structural integrity, thus making it much weaker.[5] There are problems of exclusion of parts of communities, with reaching agreement (which can result in lengthy delays) and with meeting the conditions of external funders. However, these problems also exist with state run programs. The language of recovery has evolved and "community led" is now less common than a "resilience" approach. A paradox in some resilience recovery material or guides is that even though the emphasis is on the affected community, there is an assumption that the state would provide support and capacity building.

In summary, there are a number of key organizing principles to the deployment of urban planning as key contributor to recovery governance:

- The full range of spatial planning principles are deployed, rather than focusing only upon "traditional" approaches in a place, such as regulations and policies alone.
- Accountability, transparency, and the rule of law are supported.
- Participation and acknowledgment of local knowledge and particularities are matched with responsiveness and interaction with the use and dissemination of scientific and professional evidence.
- Extraordinary agencies do not cut across routine processes that would build capability, where possible.
- Community functions and principles of fairness and equity are held as higher ideals, including efficient use of materials and resources.
- The ongoing adaptability of settlements and building of capacity are key goals, as well as making improvements to the resistance of settlements to ongoing known threats in recovery processes.
- Planning is a fundamental element in coordination, specialization, and decision making in recovery.
- Existing and future generations are considered in decision making, as are ecological, cultural, and economic decisions.
- Pre-event planning for recovery is included in planning processes, balancing a range of non-risk and other needs.

[5] See Chapter 13 for more details.

CONCLUSIONS

Urban planning, as a form of collective decision making relating to spatial outcomes, has goals oriented to individual, community, and state-oriented capacity building. At the same time, it seeks to balance these intentions against some level of centralized imposition of rules and parameters that ensure achievement of equity, efficiency, and ecologically sound outcomes across overall settlements, regions, and nations. Following from this, a fundamental tenet of recovery is that it should lead to improvements, especially regarding risk reduction over time, in the ways that the built environment is managed in terms of the range of incremental individual decisions and overarching directions for change that inform smaller scale decisions.

Recovery should not re-create risks or develop additional vulnerabilities. The often used slogan for this is "building back better." In urban planning terms it is tensioned against the challenge for retention of pre-event settings, for it is common that many aspects of a community may be in a serviceable state and which represent a significant level of investment. Accordingly, the prospect of modifying road patterns, tenure boundaries, building regulations, and ways of using land generally, particularly those that are expensive to comply with, may mean that trade-offs are made that erode the ability to improve risk levels during recovery. The underlying principle is to link post-disaster reconstruction with longer term risk reduction and mitigation to ensure that the same conditions of exposure and vulnerability are not repeated.

REFERENCES

Abercrombie, P. (1943). *Town and country planning*. London: Oxford University Press.

Alexander, D. E. (1999). *Natural disasters*. London: Kluwer.

Alliance Development Works, UNU-EHS, & The Nature Conservancy. (2012). *The world risk report* (Retrieved from Geneva).

Bavinck, M., de Klerk, L., van der Plaat, F., Ravesteijn, J., Angel, D., Arendsen, H., ... & Zuurendonk, B. (2015). Post-tsunami relocation of fisher settlements in South Asia: Evidence from the Coromandel Coast, India. *Disasters*, *39*(3), 592–609. http://dx.doi.org/10.1111/disa.12113.

Blaikie, P. M., Cannon, T., Davis, I., & Wisner, B. G. (1994). *At risk: Natural hazards, people's vulnerability, and disasters*. London, UK: Routledge.

Buxton, M., Goodman, R., & March, A. (2012). Planning systems, urban form and housing. In R. Tomlinson (Ed.), *Australia's unintended cities: The impact of housing on urban development*. Canberra: CSIRO Publishing.

Cleetus, K. (2015). *Overwhelming risk rethinking flood insurance in a world of rising seas* (Retrieved from Cambridge, MA).

Coppola, D. P. (2011). Recovery. In D. P. Coppola (Ed.), *Introduction to international disaster management* (2nd ed.). Amsterdam: Butterworth-Heinemann.

De Silva, D. A. M., & Yamao, M. (2007). Effects of the tsunami on fisheries and coastal livelihood: A case study of tsunami-ravaged southern Sri Lanka. *Disasters*, *31*(4), 386–404. http://dx.doi.org/10.1111/j.1467-7717.2007.01015.x.

Halligan, J., & Power, J. (1992). *Political managment in the 1990s*. Melbourne: Oxford University Press.
Healey, P. (1997). The revival of strategic spatial planning in Europe. In P. Healey, A. Khakee, A. Motte, & B. Needham (Eds.), *Making stretegic spatial plans*. London: UCL Press.
Hoch, C. (1994). *What planners do: Power, politics and persuasion*. Chicago: American Planning Association.
Hopkins, L. D. (2001). *Urban development: The logic of making plans*. Washington, DC: Island Press.
ISO31000. (2009). *AS/NZS ISO 31000:2009: Risk management – Principles and guidelines*.
Keeble, L. (1952). *Principles and practice of town and country planning*. London: The Estates Gazette.
Kornakova, M. (2016). *Improving the integration of urban planning with disaster risk reduction* (Ph.D.). University of Melbourne (Unpublished).
Krizek, K., Forysth, A., & Slotterback, C. S. (2009). Is there a role for evidence-based practice in urban planning and policy? *Planning Theory & Practice*, *10*(4), 459–478. http://dx.doi.org/10.1080/14649350903417241.
Levy, J. M. (2000). *Contemporary urban planning* (5th ed.). New Jersey: Prentice Hall.
Lindblom, C. E. (1965). *The intelligence of democracy: Decision making through mutual adjustment*. London: The Free Press.
March, A. (2012). *The democratic plan: Analysis and diagnosis*. Farnham: Ashgate Publishing.
McLoughlin, B. (1969). *Urban & regional planning: A systems approach*. London: Faber and Faber.
Meerow, S., Newell, J. P., & Stults, M. (2016). Defining urban resilience: A review. *Landscape and Urban Planning*, *147*, 38–49.
Meyerson, M., & Banfield, E. (1955). *Politics, planning and the public interest: The case of public housing at Chicago*. New York: Free Press.
Mumford, L. (1968). *The urban prospect: The highway and the city*. New York City: Harcourt Brace Jovanovich.
Sandercock, L., & Kliger, B. (1998). Multiculturalism and the planning system. *Australian Planner*, *35*(No.3 & No.4) 127–132 & 223–227.
Taylor, N. (1998). *Urban planning theory since 1945*. London: Sage.
Tierney, K. (2012). Disaster governance: Social, political, and economic dimensions. *Annual Review of Environment and Resources*, *37*, 341–363.
UNISDR. (2015a). *Global assessment report on disaster risk reduction. Making development sustainable: The future of disaster risk management*. Retrieved from: http://www.preventionweb.net/english/hyogo/gar/2015/en/gar-pdf/GAR2015_EN.pdf.
UNISDR. (2015b). *Reconstructing after disasters: Build back better. Ministerial roundtable*. Sendai: UN World Conference on DRR, 3.
UNISDR. (2015c). *Sendai Framework for Disaster Risk Reduction 2015–2030. Presented at the 3rd United Nations World Conference on DRR Sendai, Japan*. Retrieved from: http://www.unisdr.org/we/coordinate/hfa-post2015.
Wilkinson, E., Comba, E., & Peters, K. (2014). *Disaster risk governance: Unlocking progress and reducing risk*. UNDP: Geneva, Switzerland. Retrieved from: http://www.undp.org/content/undp/en/home/librarypage/crisis-prevention-and-recovery/disaster-risk-governance-unlocking-progress-and-reducing-risk.html.
Yiftachel, O. (1999). Planning theory at a crossroad: Tthe third Oxford conference. *Journal of Planning Education and Research*, *18*, 267–270.

FURTHER READING

Coke, J. G. (1968). Antecedents of local planning. In W. I. Goodman, & E. C. Freund (Eds.), *Principles and practice of planning* (4th ed.) (pp. 7–28). Washington: International City Managers' Association.

CHAPTER

Equity in Recovery

3

Janet Stanley[1,2]

[1]*The University of Melbourne, Melbourne, VIC, Australia;* [2]*National Centre for Research in Bushfire and Arson, Melbourne, VIC, Australia*

INTRODUCTION

This chapter reports on equity as an aspect of the capacity to recover from disasters. The concept of equity is explored, as well as the nature of disasters and a discussion about the ability to recover. People and communities, both in developed and developing countries, do not have equal recovery capacity. Recovery depends on a range of factors including the type of the disaster, personal resources, the local community, and government planning, particularly in relation to predisaster planning. This is not just response planning but planning to place the community in the fittest condition prior to the next potential disaster and to strengthen the resilience of those experiencing disadvantage.

While there are many types of disasters, this chapter concentrates on large-scale environmental disasters as the frequency of occurrence and the size of their impact on people are growing. This is due to the increase in world population, particularly population that is concentrated in urban areas, and the increase in extreme events leading to the risk of disasters arising due to climate change. The bulk of population growth is occurring in Asia. Between 1980 and 2010, cities in the Asia–Pacific region grew by around one billion people and United Nations projections show that a further one billion will be added by 2040 (UN Habitat, 2015). More than 75% of the increase in urban cover is projected to be in Asia, a trend expected to last for decades to come (Walsh, 2012). Unfortunately, many regions are unprepared, and without the capacity to assimilate this growth within short time frames, they have to be prepared for any impending disasters (Dávila, 2013). In contrast to most developed countries, population growth in Australia is exceptionally large, with also an accompanying trend toward growth in urban areas, including low-density growth along coastal strips and penetration into forested areas on the fringe of cities. This growth is posing many challenges for planning in general and for disaster recovery.

The growth of populations, in both developing and developed countries, increases the likelihood of individuals being located in places where disasters are more likely to occur. This is especially so with urban development penetrating into at-risk areas, such as on low-lying land (which may have been drained) and in forest areas, as is

Urban Planning for Disaster Recovery. http://dx.doi.org/10.1016/B978-0-12-804276-2.00003-7

the case on the edge of major cities in Australia. Peat forests are being drained and the forests are cleared for agriculture, especially palm oil, in Indonesia, thus increasing the fire risk of the now dry, underground peat. Urban areas are penetrating into forested areas in Australia, Greece, Spain, and parts of the United States, thus again increasing the bushfire risks and occurrences.

DISASTER EVENTS

While the potential for catastrophic events has always been present, anthropogenic climate change is resulting in the rise of extreme events, thus exacerbating the risk of environmentally initiated disasters. The growth of greenhouse gases in the atmosphere is leading to rising temperatures, including rising sea levels, changes in rainfall patterns, ocean acidification, and storm and cyclone events. These events are increasing the risk of disasters, such as bushfire, flood, storm and cyclone damage, and prolonged high temperatures. An Intergovernmental Panel on Climate Change (IPCC) report on extreme events notes that:

> *A changing climate leads to changes in the frequency, intensity, spatial extent, duration, and timing of extreme weather and climate events, and can result in unprecedented extreme weather*
>
> **Allen et al. (2011, p. 5)**

A natural disaster may also arise due to the length, repetition, and cooccurrence of climate events. For example, many parts of Asia (at the time of writing, May 2016) are subject to an extended drought. This event is currently encompassing New Guinea, Vietnam, Burma, and India (McDonald, 2016, p. 12). While each extreme climate event may result in a disaster, the combination of events, such as rising sea levels and storm activity, is also likely to compound the adverse impact on people and the environment, challenging resilience and a good recovery.

Repeated events were seen in Queensland, where severe flooding occurred over an extended time. The rain in December 2010 came after a wet spring and caused nine floods that affected almost 1,300,000 sq. km of land, caused billions of dollars in damage, led to the evacuation of thousands of people, and resulted in 35 deaths (University of NSW, 2012). Severe flooding and Cyclone Oswald occurred in late January 2013, flood waters peaking at 9.53 m in the town of Bundaberg, accompanied by a series of tornadoes (Daily News, 2013). Four deaths were recorded. The 2010–11 Queensland floods were attributed to a La Niña event that brought very heavy rain to the east coast of Australia. Work by the Bureau of Meteorology (2012) has shown that record high sea surface temperatures in October to December 2010 also contributed to the record rainfall.

The major reinsurance company, Munich Re, has documented the world trend in natural disasters (2016). While there is a fairly stable pattern of geophysical events from 1980 to 2015, there is a steady increase in other environmental disasters (storms, floods, drought, and fire) (Fig. 3.1). In addition to the increase in the number

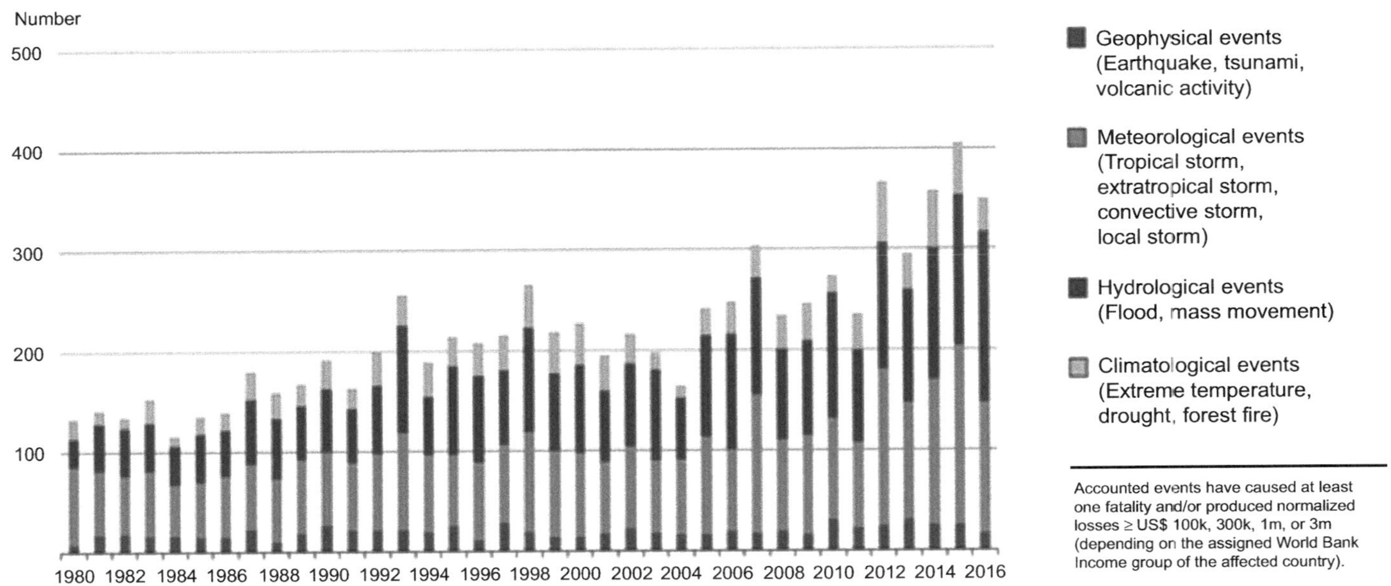

FIGURE 3.1 World Environmental Disasters From 1980 to First Half of 2016.

Munich Re (2016)

of events, their data reveals an increase in the impact of these events when the past 30 years is compared with the last 10 years (Höppe, 2015). Munich Re defines a catastrophic event as one which results in a direct insured loss to properties of US $25 million or more (2014 values). However, this definition is likely to underestimate the number and severity of events. In developing countries the rate of insurance uptake is less than in developed countries, but insurance uptake in a developed country may also not be high, with 30% of homes in the Victorian 2009 bushfire having no insurance cover. Floods resulting from a hurricane are not counted as this is not covered by insurance, and insurance does not include events that don't involve property loss, such as the impact on people of a prolonged heat wave.

In the counted events, Munich Re measures the number of fatalities, overall losses, and insured losses. Thus, many of the impacts of a disaster remain uncounted, such as the number of physical injuries, which are often high, even in a developed country. In the first 72 h of the February 7, 2009, bushfires in Victoria, 414 people presented to hospitals, as a result of the fires (Cameron et al., 2009). Psychological injuries and stress reactions remain uncounted, as does loss of business revenue and the actual businesses. Indeed, many losses from disasters are not accounted for in many sources that estimate the cost of disasters. Also rarely mentioned is the reality that poorer people disproportionately experience natural disasters.

EQUITY

Inequality describes differences between people in terms of income and wealth, as well as education, health, and other social outcomes. A lack of equity implies unfairness and that it is preventable (Douglas, Friel, Denniss, & Morawetz, 2014). As Stiglitz says (2012), inequity is a choice. Inequity is present and growing within many countries. For example, referring to Australia, the authors note that:

> *In recent decades the income share of the top 1 per cent has doubled, and the wealth share of the top 0.001 per cent has more than tripled. At the same time, poverty is increasing and many of those dependent upon government benefits, including the unemployment benefit, have fallen well below the poverty line*
>
> **Douglas et al. (2014, p. 8)**

Many countries reflect a similar pattern of inequity with the discrepancy in wealth higher than the discrepancy in income (Dabla-Norris, Kochhar, Suphaphiphat, Ricka, & Tsounta, 2015). Inequity entails more than financial issues. Inequity in health care access and use is more pervasive in developing countries. Women are disproportionately represented in the lowest income groups, and low-income earners are concentrated in certain regions, often magnifying the negative effects of inequity and creating clusters of socioeconomic deprivation (Demalo, 2014).

A common perspective to correct inequity is that all people should have equal opportunities. Sen (1992) uses the word "capabilities" to describe this, arguing that all people should have equal capabilities to achieve what they wish. This perspective is often reflected in minimum standards, such as the provision of welfare benefits by many governments. In the case of disaster recovery, an example could be the ability to obtain a minimum standard of postdisaster shelter for all people, perhaps through low or no-interest loans to those who are left with few resources after the disaster. A capability approach has the advantage that it offers an equal platform that enables people to make their own choices about recovery, thus maintaining their own decision making and control over their future.

ABILITY TO RECOVER FROM A DISASTER

The ability to recover from a natural disaster will depend not only on the nature of the disaster but also, in large part, on the predisaster conditions. This includes personal resources, government planning, and resources available, as well as local conditions, such as community resources and social capital, the latter also being influenced in large part by government policy and planning. Those who have preexisting disadvantage or low resources are likely to be less able to recover from a natural disaster than those with more resources and better health and well-being. This suggests that two factors are important in disaster recovery. There is a need to improve the capabilities of those with the least resources to maximize their ability to recover after a disaster and also give priority to those who remain with the lowest capabilities postdisaster in the recovery stage.

NATURE OF THE DISASTER

The characteristics of natural disasters is such that there may be a need to respond to an immediate unpredictable event, such as bushfire, or there may be time to prepare for a more predictable outcome, such as longer term water shortages in Southeast Australia. Of course, as in the latter, even while an event may be more predictable, clarity of details may be less clear when a longer time span is involved. Thus, an event can placed anywhere on the two dimensions illustrated by the stars in Fig. 3.2.

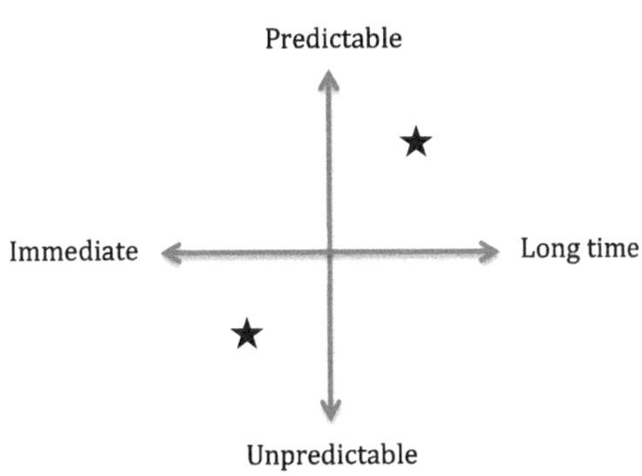

FIGURE 3.2 The Two Dimensions of a Disaster.

Even less certain will be the indirect effects associated with a disaster and thus the ability to recovery from these impacts. For example, where a disaster, such as flooding or a cyclone, occurs in a major food growing area, this may limit food supplies for a populated area or the commercial viability of a township in a region that processes the farm products. A further indirect impact may occur in that the shortage of certain foods may push up the price of these foods, thus placing their availability beyond the reach of some groups in the community. An event may carry some longer term impacts on health, such as the outbreak of cholera, or contamination of the water supply. Infrastructure damage on roads and buildings may need a longer timeline for repairs.

Severe, long-term indirect impacts occurred in 2010, when Pakistan experienced the worst floods in its history with one-fifth of the country flooded, effecting 15 million people (Kron, 2014). Over 1760 people were killed, drinking water was contaminated causing illness, and, significantly, 200 hospitals and medical centers were also flooded. Food supplies and livelihoods were washed away and 1.5 million homes were destroyed or damaged. Thus, the nature of the event (type, severity, location) has impacts on the ability to recover. Repeated or ongoing exposure to severe events risks a reduced capacity to recover, in environmental, human, and economic terms. The current drought in Asia (at the time of writing, May 2016) is leading to extensive consequences, including rice production being badly hit and severe water shortages, all of which will delay the recovery process, especially when the consequences are widespread, where personal and government resources are low in the first place and there is a risk of conflict occurring when recovery is difficult due to low levels of resources (McDonald, 2016, p. 12).

PERSONAL RESOURCES

As noted earlier, the impact of a natural disaster on households and communities will depend on their access to resources. In many situations people will adjust to adverse unintended impacts and absorb the cost, however, for some people and some communities, even a more minor event may result in long-term adverse outcomes. For example, after the Kinglake fires (in Victoria) in 2009, some people lacked the financial resources to rebuild their homes and now occupy government housing in a highly disadvantaged location in Melbourne; others have become homeless (personal communication). Personal factors that influence recovery include knowledge and information about how to prepare for, and respond to, the nature of the disaster and about personal, social, psychological, and economic resources available to undertake this response.

Those who are already experiencing disadvantage and social exclusion are more likely to have the most difficulties in both the short-term response to a disaster and the long-term process of recovery (Stanley, 2014). Their position of exclusion is commonly associated with fewer resources to deal with challenges. A second group of people who are "just managing" prior to the disaster may be at risk of being moved into poverty and poorer well-being post disaster. Thus, some groups and some locations will need particular attention given to them in predisaster planning and assistance, and disaster response and recovery, as identified below.

THOSE WITH EXISTING VULNERABILITIES

People who need special consideration are those with existing vulnerabilities, who have:

- a low income and low wealth, thus little choice and flexibility, and lack of credit to negotiate loans;
- poor access to knowledge and poor ability/experience/power/connections in negotiating better outcomes with government and bureaucracy; and
- a lack of, or limited, social capital to call upon to provide support and assistance. This may be due to a low capacity to make and maintain social networks, such as due to long working hours or mental illness, or it may be due to geographical isolation from support networks and services.

In the developing world there remains many countries where the majority or close to the majority of people are represented in the first two categories of vulnerability, particularly in African countries and also in places such as India, Pakistan, Bangladesh, and Laos (Roser, 2016).

Following the floods in Queensland in 2010–11, the Queensland Council of Social Services found that those already experiencing disadvantage were disproportionately adversely impacted (ACOSS, 2013). Particular areas included:

- lack of insurance, or underinsurance, and the rejection of flood insurance claims left people unable to live in or repair their homes;
- loss of employment through disruptions to and closure of local businesses;
- loss of rental tenancies and inability to meet higher bond payments and rents;
- increased pressure on public housing waiting lists; and
- increased living costs.

With over 105,000 people officially defined as homeless on any given night in Australia, those who are already experiencing homelessness find disaster recovery difficult (Pendrey, Carey, & Stanley, 2012). This study supported the claim that a natural disaster tends to magnify preexisting disadvantages and health issues for homeless people and those in insecure accommodation, particularly around access to safe shelter, fresh water, transport, along with problems of mental illness, chronic disease, substance abuse, and posttraumatic stress. The type of natural disaster manifests particular difficulties. For example, in a flood situation, those experiencing disadvantage may be living in less satisfactory accommodation, such as caravan parks, which tend to be located in flood prone areas. Unable to locate elsewhere, people remained in wet conditions for an extended time. The study in Victoria found that living in a flood situation increased vulnerability to insect bites and skin infections and people found it difficult to keep clothing dry. The demise of public transport after a disaster renders it more difficult for homeless people to access resources or move out of the area. Communication is difficult with those who are homeless because of their lack of phones and a fixed address.

THOSE WHO ARE AT RISK OF BECOMING VULNERABLE

A second group who may need special attention due to their potential vulnerability are those who were only just managing prior to the disaster (Stanley, 2014). An example here might be a farming household where all resources are used to maintain the farm prior to the extra demands from a disaster, which may subsequently lead to a further reduction in farming viability. Businesses and residences adversely impacted by a disaster, such as bushfire, may find that the price on sale drops considerably. Low-income home owners may not have taken out insurance (30% did not have insurance in the 2009 Victorian bushfires), or are underinsured, thus find that they are unable to rebuild. Indeed, the numbers of people in this category are potentially quite large in both developed and developing countries. It includes those who are frailer, have chronic illness, disability, or mental illness.

Often the indirect impacts arising from the disaster, discussed above, compound the difficulties in the recovery process for many people without the resources to support themselves. For example, the disaster may well disable infrastructure and services, leaving roads damaged and government offices and services, such as schools and hospitals, unusable. Damage to housing will reduce the availability of rental stock within, and nearby, the disaster area, as well as increase the waiting lists for government housing. Prices are likely to rise for available housing, and other necessities and local sources of assistance, such as community services, may become overwhelmed with demands. The Climate Institute reports that disturbed behavior is more common after an extreme event, such as substance abuse, family violence, and self-harm, and suicides rise as much as 8% (Doherty & Clayton, 2011). Both children and adults may suffer posttraumatic stress and lingering behavioral issues associated with fear or anxiety.

SOCIETAL RESPONSES

BROAD SOCIETY

Making decisions about the social justice position of society is difficult, as illustrated in the disputes around the IPCC meetings, where vested interests often prevail. Disasters in developing countries, which have a smaller economy and lower resources, commonly require considerable assistance from the United Nations and international agencies, as well as wealthier countries, to manage disaster recovery. This assistance is often short-term and unfortunately long-term recovery prospects are usually less certain. Nepal experienced a 7.8-magnitude earthquake on April 25, 2015, that killed 9000 people, injured 22,000 people, and damaged or destroyed 800,000 homes. One year later, little reconstruction has taken place (Taylor, 2016). Attention to equity within countries in response to a disaster is often left to charitable organizations, subsidized by government, but still heavily reliant on donations from the public, as largely the case in Australia.

As noted, disasters are increasingly being linked to greenhouse gas emissions, leading to climate change and more extreme events, with the risk of disasters. Developed countries generate the greatest greenhouse gases, particularly with their dependence on fossil fuels. Within a country, the highest emitters of greenhouse gases are the wealthiest people. A very poor household has emissions of about 22.3 tons annually; households with one person working, who earns an average wage, emits about 28.6 tons annually; whereas, a high-income, tertiary-educated household emits on average 57.8 tons annually (NIEIR, 2007). Such evidence strengthens the social justice argument for equity where disasters are linked with climate change, as those who have the greatest culpability in creating the problems should pay the greatest amounts for the impacts.

Equity can also be argued on pragmatic grounds. A society with little disparity between people is a society where all people are better off (Wilson and Pickett, 2010). The International Monetary Fund reports that an increase in the income share of the lowest 20% of income earners is associated with higher growth in the country's Gross Domestic Product (Dabla-Norris et al., 2015). These findings are based on a wide range of areas where inequity can occur, reflected in disparity in child development, health, levels of crime, obesity, trust, and mental illness.

THE THIRD SECTOR

A disaster also adversely impacts on the ability of an agency to provide services, including meeting the increase in demand for assistance and the need to divert resources away from usual clients to assist people who would not be a client except for the emergency (Pendrey et al., 2012). Pendrey et al. (2012) found that only a small number of agencies had incorporated disaster risk into their organizational planning. Many were underinsured. Most community service agencies do not have the resources to respond to emergency events, especially when they occur every year or every few years, as has been the trend with the recent floods and bushfires in Australia or the extensive size of events as illustrated above in Pakistan. This is despite the fact that these agencies are often called on by government both to build community resilience and to respond to events.

COMMUNITY

Social inclusion or exclusion broadens the ideas of barriers to full participation in society beyond this being only a question of income, to being a multidimensional concept, including employment, participation, support, and political activity, as well as income. Fig. 3.3 shows the most important conditions (1% statistical significance) derived from over 1000 personal interviews with a stratified random sample undertaken in Victoria, Australia (Stanley, Stanley, & Hensher, 2012). The statistical modeling showed that a person who has higher social inclusion is also likely to have

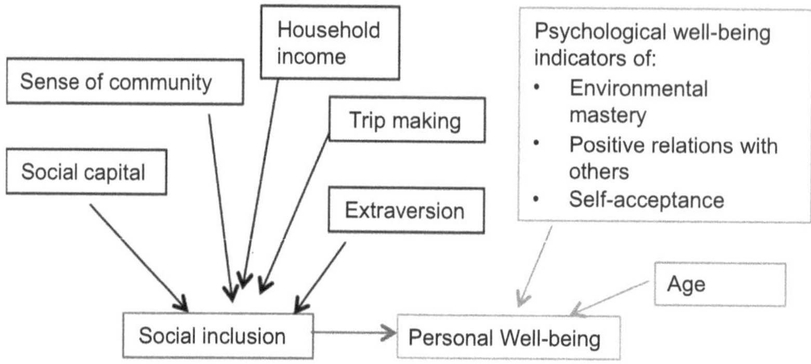

FIGURE 3.3 Drivers of Social Exclusion and Well-Being.

higher levels of personal well-being. In addition to an adequate income and having an extraverted personality, social inclusion is built from social capital, attachment to community, and the ability to be mobile.

Thus, improving capabilities is not only achieved through monetary handouts but also through measures that increase social capital, connections to the community, and the ability to travel. Improving opportunity, participation, and inclusion will increase personal resources and the ability of people to recover from a natural disaster (Stanley & Vella-Brodrick, 2011).

In July 1995 a heat wave in Chicago killed 739 people (Klinenberg, 2013). It was found that those in poorer neighborhoods were more vulnerable to the heat, but not in all poor neighborhoods. Those in poor neighborhoods who fared best came from places that had sidewalks, stores, restaurants, and community organizations that brought people together. During the extreme heat, people in this neighborhood were doing "wellness" checks on others, knocking on doors, and the community knew who was vulnerable. Living in this neighborhood with strong social capital prolonged life expectancy such that it was found to be roughly equivalent to having an air conditioner in each room!

The ability to work together facilitates active and social learning. This encompasses trial and error where communities generate knowledge and obtain new knowledge and adjust decisions (Berke, 2002). This is very important as responding to climate change, extreme events, and the possibility of disaster is not a "once-off" event. In a context of uncertainty and few ready answers, people need to continually adapt, respond, and change. People with experience and skill who have learned and are confident in this process will be invaluable to guiding others through this continually evolving process, thus improving the confidence and resilience of a community.

GOVERNMENT PLANNING

In large part, the capacity for recovery after a disaster is dependent on the predisaster planning that has taken place, particularly the quality and comprehensiveness of this planning (Berke, Cooper, Aminto, Grabich, & Horney, 2014). Berke et al. (2014, p. 315) reviewed 87 disaster recovery plans in the United States and found that the plans have a "weak framework to guide recovery decisions to achieve long-term resiliency."

It is argued in this chapter that comprehensive recovery is in part dependent on predisaster planning for social outcomes, with particular attention being given to those people and places with known vulnerabilities. Government planning is undergoing considerable change at present. Chapin (2012) outlines four waves of changes undergone by planning in the United States since 1950. He describes the last emerging wave as one that will be dominated by sustainable growth, which includes new policy areas such as site and neighborhood design, climate change, and change that will "tear down the institutional and intellectual silos that have limited effectiveness…" (Chapin, 2012, p. 11). This new direction also needs to encompass a broader uptake of social planning, such as the concept of "complete communities" adopted in Vancouver, where there is a strong integration of social outcomes in housing, transport, and health, all important components of equality, social inclusion, and well-being (Ohland & Brooks, 2013). How to achieve this integrated governance across functional divisions of government departments is a major challenge for planning.

As argued earlier in this chapter, priority attention should be given to particular issues associated with those people at risk of social exclusion. For example, land at greater risk of flooding, fire, or inundation by the sea is often offered at a lower price and thus purchased by people with a lower income, increasing their vulnerability. Those on a lower income are less likely to use building materials and designs that offer greater protection from extreme events as they are often more expensive than more conventional materials. Setting priorities for attention is difficult. The following is a suggested guideline for planning priorities that identifies and attends to those with the greatest needs:

Priority 1: Those at risk of social exclusion or poor well-being who have other vulnerabilities, who live in a location with a higher vulnerability to disaster, which has poor infrastructure and/or poor preparation for a disaster event. An example could be a lone parent on a low income who lives in an area with high risk of fire with poor transport services.

Priority 2: Those at risk of social exclusion or poor well-being with other vulnerabilities who live in a location with a greater disaster risk, for example, a lone parent on a low income who lives in an area with high risk of fire.

Priority 3: Those at risk of social exclusion or poor well-being with more than one vulnerability, for example, a lone parent on a low income.

Priority 4: Those at risk of social exclusion or low well-being, such as a lone parent.

Urban design can be undertaken to reduce the impact of disasters. For example, access to public cool areas, planting of trees to reduce the urban heat island impact, and the provision of water features to lower temperature can counter heat stress, especially for people who are homeless or have low-quality housing with little insulation.

With the higher incidences of disasters associated with increasing greenhouse gas emissions, there is likely to be a need to change some planning legislation to facilitate response to these changes. For example, planning legislation in Victoria relating to fire management covers buildings and the immediate area of development. It fails to allow for consideration of the additional fire risks due to climate change and take account of the wider environmental location of a building development (Stanley, 2015).

A further challenge for planning is how to effectively combine vertical governance or decision making from the bottom up with decision making from the top down, in order to integrate local citizen participation with broader, strategic planning goals. Public engagement to create a disaster plan that reflects local values, needs, and capabilities was one of six quality principles which arose from Berke's et al. (2014) study of disaster planning in the United States. Participation increases the likelihood that a neighborhood will be structured along the lines desired by local residents, who are often familiar with the local strengths and resource gaps. Such an approach will also build buy-in to community decisions and ownership of recovery plans. Indeed, community participation in decision making builds a sense of community, social capital, and engagement, as well as an opportunity to grow capabilities, self-esteem, and confidence, along with leadership possibilities. For example, a system can be designed where people are designated to check on vulnerable people during an extreme heat event, such as those who are house bound or elderly can be put in place.

It should be noted that the preceding points largely refer to disaster planning in developed countries. Disaster planning in developing countries needs considerable more attention because of the size and frequency of natural disasters and the extent of inequity present in many countries. The challenge is even greater as there is often an absence of sufficiently trained planners who adopt the contemporary thinking about inclusion and environmental sustainability, both critical variables for disaster recovery planning (Lehmann & Thornton, 2015).

CONCLUSIONS

The growing impact of climate change suggests that the trend of increasing natural disasters will continue. When these risks are aligned with increasing urban population growth and the growth in inequity, then the task of planning for disaster recovery becomes more urgent. Learnings from previous disasters reveal that those people with the highest predisaster vulnerabilities and fewer personal and community resources have the greatest difficulties recovering from disasters. There is a risk that some will not recover on a long-term basis with regard to employment, housing, and health problems. Thus, this chapter argues that recovery plans should particularly target those people and locations that have particular vulnerabilities. Predisaster planning should address the particular issues of vulnerability, such as avoiding accommodation being located on land subject to the risk of flooding and building personal capabilities and community strengths and supports.

The responsibility for successful recovery planning involves all levels of society, government, communities, not-for-profit organizations, and business. The case for involvement still needs to be argued, with a risk that international commitment for assistance not eventuating and the capacities of not-for-profit organizations being stretched too far. Government at the more local level has the challenging task of integrating across the functional areas, such as transport, health, and housing, while also coordinating governance, tasks, and actions, as well as offering information, leadership, and support across the community. The good part of this challenge is that where the community is offered this encouragement, supported by adequate resources, much of this planning and facilitation of action will be undertaken by a willing community, as evidenced in Australian research (Stanley et al. 2013). This study showed that given the opportunity, most communities are willing and able to make decisions and support other community members where the need arises.

REFERENCES

(ACOSS) Australian Council of Social Services. (January 2013). *Extreme weather, climate change and the community sector: ACOSS submission to the senate inquiry into recent trends in and preparedness of extreme weather events.* ACOSS, ACT.

Allen., et al. (2011). *Managing the risks of extreme events and disasters to advance climate change adaptation: Summary for policy makers* (Special report of the intergovernmental panel on climate change). Retrieved from: https://www.ipcc.ch/pdf/special-reports/srex/SREX_FD_SPM_final.pdf.

Berke, P. R. (2002). Does sustainable development offer a new direction for planning? Challenges for the twenty-first century. *Journal of Planning Literature*, *17*(1), 21–36.

Berke, P., Cooper, J., Aminto, M., Grabich, S., & Horney, J. (2014). Adaptive planning for disaster recovery and resiliency: An evaluation of 87 local recovery plans in eight states. *Journal of the American Planning Association*, *80*(4), 310–323.

Bureau of Meteorology, Record Breaking La Niña Events. July 2012, Australian Government.

Cameron., et al. (2009). Black Saturday: The immediate impact of the February 2009 bushfires in Victoria, Australia. *The Medical Journal of Australia*, *191*(1), 11–16.

Chapin, T. (2012). Introduction. *Journal of the American Planning Association*, *78*(1), 5–15.

Dabla-Norris, E., Kochhar, K., Suphaphiphat, N., Ricka, & Tsounta, E. (June 2015). *Causes and consequences of income inequality: A global perspective.* International Monetary Fund.

Daily News. (February 3, 2013). *Tropical cyclone oswald causes major flooding in Queensland, Australia.* Retrieved from: www.nydailynews.com/news/australia-flooding-gallery-1.1251124#ixzz2JmPWTnRG.

Dávila, J. D. (2013). *Urban mobility and poverty: Lessons from Medellin and Soacha, Colombia.* Retrieved from: http://discovery.ucl.ac.uk/1366633/.

Demalo, A. (August 28, 2014). *Social inequality: Is the elephant in the room strangling us all?* The Conversation. Retrieved from: http://theconversation.com/social-inequality-is-the-elephant-in-the-room-strangling-us-all-30574.

Doherty, T., & Clayton, S. (2011). The psychological impacts of global climate change. *American Psychologist*, *66*(4), 265–276.

Douglas, B., Friel, S., Denniss, R., & Morawetz, D. (May 2014). *Advance Australia fair? What to do about growing inequality in Australia*. The Australia Institute.

Höppe, P. (2015). *Update on El Niño & global natural catastrophes, natural catastrophe review webinar*. Munich Re. Retrieved from: https://www.munichre.com/site/mram-mobile/get/documents_E336591247/mram/assetpool.mr_america/PDFs/4_Events/MunichRe_III_NatCatWebinar_071415_pdf.pdf.

Klinenberg, E. (January 7, 2013). *Adaptation: How can cities be climate-proofed?* The New Yorker. Retrieved from: https://ipk.nyu.edu/publications/item/505-klinenberg-adapatation.

Kron, W. (October 17, 2014). *Flood disasters – A global perspective*. Munich: Munich Reinsurance. http://wp.iwaponline.com/content/17/S1/6.

Lehmann, S., & Thornton, K. (2015). Looking Ahead: Applying low carbon principles to shift urban design paradigms in the Asia–Pacific region towards green urbanism. In S. Lehmann (Ed.), *Low carbon cities: Transforming urban systems* (pp. 394–421). London: Routledge.

McDonald, H. (May 21–27, 2016). *Asia's drought crisis may lead to water wars*. The Saturday Paper.

Munich Re. (2016). *World, natural catastrophes*. Insurance Information Institute. Retrieved from: http://www.iii.org/fact-statistic/catastrophes-global.

NIEIR (National Institute of Economic and Industry Research). (2007). *The impact of carbon prices on Victorian and Australian households*. NIEIR (National Institute of Economic and Industry Research) 2007 (A report for the brotherhood of St Laurence, Melbourne).

Ohland, G., & Brooks, A. (2013). *Are we there yet? Creating complete communities for 21st century America*. Washington, DC: Reconnecting America. Retrieved from: http://www.ssti.us/wp/wp-content/uploads/2012/10/20121001AreWeThereYet-web.pdf.

Pendrey, C., Carey, M., & Stanley, J. (March 2012). *Climate change, extreme weather and the health and wellbeing of people who are homeless: Report to participants*. Monash Sustainability Institute.

Roser, M. (2016). *World poverty, our world in poverty*. Retrieved from: https://ourworldindata.org/world-poverty/#share-of-the-population-below-the-international-poverty-line.

Sen, A. (1992). *Inequality re-examined*. London: Oxford University Press.

Stanley, J. R. (2014). Climate change: A new challenge for social policy. In A. McClelland, & P. Smyth (Eds.), *Social policy in Australia: Understanding for action* (3rd ed.). Australia: Oxford University Press.

Stanley, J. (March 2015). *Fire and the proposed land use change at Arthurs Seat*. Peninsula Preservation Group Inc. (Unpublished).

Stanley, J., Birrell, B., Brain, P., Carey, M., Duffy, M., Ferraro, S., … & Wright, W. (2013). *What would a climate-adapted settlement look like in 2030? A case study of Inverloch and Sandy point* (Report for the National Climate Change Adaptation Research Facility, Gold Coast).

Stanley, J. K., Stanley, J. R., & Hensher, D. (December 2012). Mobility, social capital and sense of community: What value? *Urban Studies*, *49*(16), 3595–3609.

Stanley, J., & Vella-Brodrick, D. (2011). What leads to social inclusion? An examination of trips, social capital and well-being. In G. Currie (Ed.), *New perspective and methods in transport and social exclusion research* (pp. 187–200). UK: Emerald Press.

Stiglitz, J. (2012). *The price of inequality*. London: Allen Lane, The penguin Group.

Taylor, A. (April 25, 2016). *Nepal's earthquakes: One year later*. The Atlantic. Retrieved from: http://www.theatlantic.com/photo/2016/04/nepals-earthquakes-one-year-later/479772.

UN Habitat. (2015). *The state of Asian and Pacific cities 2015. Urban transformations shifting from quantity to quality*. United Nations Human Settlements Programme (UN-Habitat).

University of NSW. (May 2012). *Triple Whammy: Ocean warming, La Niña, & cyclone produced Queensland floods*. Retrieved from: http://newsroom.unsw.edu.au/news/science/triple-whammy-ocean-warming-la-nina-cyclone-produced-queensland-floods/.

Walsh, B. (2012). *Urban Planet: How growing cities will wreck the environment unless we build them right*. Time Magazine. Retrieved from: http://science.time.com/2012/09/18/urban-planet-how-growing-cities-will-wreck-the-environment-unless-we-build-them-right/.

Wilson, R., & Pickett, K. (2010). *Why equality is better for everyone*. UK: Penguin Books.

CHAPTER

4

When Systems Break Down: The Role of International Aid and Humanitarian Response in Disaster Recovery

Thomas Bamforth[a]

International Federation of Red Cross/Red Crescent Societies (IFRC), Melbourne, VIC, Australia

INTRODUCTION

This chapter provides an overview of the global humanitarian system for emergency management in shelter and housing. It is based on the author's experiences as a humanitarian practitioner with the Shelter Cluster and reflects issues and concerns from a field perspective. The chapter argues that urban vulnerability is poorly understood and that natural disasters, in particular, are events that expose the social, economic, and planning fault lines that underpin poorly managed urban growth. This is done through analyzing six components of humanitarian shelter coordination (relationships with government, local civil society and the private sector, protection issues, human rights in housing and property, cash and market mechanisms in humanitarian response, and accountability to beneficiaries). Natural disasters are also opportunities to address underlying urban planning issues and vulnerabilities, but these opportunities are frequently undermined through a slow transition process from response to recovery. While humanitarian agencies are capable of adjusting to urban crises, and there are some development agencies such as UN-Habitat and Habitat for Humanity that have an urban focus to their work, these skills are not systemic to the shelter sector. There is a general lack of connection between humanitarian agencies and longer term reconstruction organizations (such as the World Bank), and weak national governance structures often undermine recovery in the medium and long term.

In managing emergency response in shelter and housing, much is dependent on context. While systems for international emergency response are relatively simple in

[a]Tom Bamforth is Global Focal Point for Shelter (Coordination) with the Global Shelter Cluster, supported by International Federation of Red Cross/Red Crescent Societies (IFRC). The views expressed in this chapter are his own.

Urban Planning for Disaster Recovery. http://dx.doi.org/10.1016/B978-0-12-804276-2.00004-9

theory, in practice they must accommodate diverse human experiences and coping strategies after disasters. In the shelter and housing sectors, emergency interventions must also assist in catalyzing early recovery for urban and rural populations, men and women, economic centers, and household livelihoods, inter alia. While taking into consideration individual and households needs separately, emergency response must occur at scale, with speed, and provide the immediate basis on which longer term recovery can occur—these are vital, yet often contrary, impulses that provide challenges for both the humanitarian system and the emergency response.

As the world moves further into the Anthropocene, an increasingly unpredictable climate and a global population that is now predominantly urban underline the importance of urban disaster preparedness, risk reduction, and response. Fifty-four percent of the current global population now live in cities, and this proportion is projected to rise to 66% by 2050 (United Nations Department of Economic and Social Affairs, 2014). Despite this, most international nongovernmental organizations (NGOs) and United Nations Agencies have their origins in agricultural development, rural emergencies, or conflict response that largely characterized the postwar boom in international aid and development. There are relatively few agencies that specialize in urban development or humanitarian response.

HUMANITARIAN REFORM AND THE TRANSFORMATIVE AGENDA

Several attempts have been made over the last decade to reform the international humanitarian system. The scale and complexity of the major international response to the Asian tsunami in 2004, which killed 230,000 people and affected many more across 14 countries, led to major calls for reform focusing on humanitarian financing, coordination of humanitarian response, and United Nations leadership (Save the Children, 2014). A reform process initiated by the United Nations Emergency Relief Coordinator, together with global civil society representatives, who form the Inter-Agency Standing Committee (IASC)—a body that provides oversight and strategic direction of humanitarian response on behalf of the UN General Assembly—sought to improve the effectiveness of humanitarian response through greater predictability, accountability, coordination, and partnership. There were three main components of humanitarian reform that now provide the institutional basis for disaster management. These are as follows:

- The *cluster approach*: addressing the need for "adequate capacity and predictable leadership in all sectors" of humanitarian response.
- *Humanitarian financing*: addressing the need for "adequate, timely, and flexible financing" of humanitarian response, notably through the Central Emergency Relief Fund.
- *Humanitarian Coordinator strengthening*: addressing the need for "effective leadership and coordination in emergencies" by the senior UN figure in country (OCHA, 2016).

Essential to the process of reform is the principle that effective humanitarian leadership is inclusive, acknowledging the significant growth in number, funding, and influence of international NGOs over the past decade, as well as the relative weakness of UN Agencies in providing overall sectorial leadership. The principle of partnership acknowledged the importance of the three main pillars response acting together: UN Agencies, NGOs, and the Red Cross/Red Crescent Movement. Commitment to partnership between these pillars was endorsed through a set of principles developed in 2007 (Global Humanitarian Partnership, 2007).

The major structural change to the humanitarian system was the introduction of "clusters" as a formal mechanism to replace the previously ad hoc and voluntaristic "sectorial" approach that had proved deficient in managing larger scale responses. The core elements of humanitarian response were divided into 11 areas relating to life-saving response foci, each with its own separate lead agency that has designated responsibility for coordinating the cluster in emergencies. The 11 IASC clusters and lead agencies are presented in the following table:

Global Humanitarian Cluster Lead Agencies

Cluster	Lead Agency
Camp coordination and camp management	UNHCR/IOM
Early recovery	UNDP
Education	UNICEF/Save the Children
Emergency telecommunications	WFP
Food security	WFP
Health	WHO
Logistics	WFP
Nutrition	UNICEF
Protection	UNHCR
Sanitation, water, and hygiene	UNICEF
Shelter	UNHCR/IFRC

IFRC, *International Federation of Red Cross/Red Crescent Societies;* IOM, *International Organization for Migration;* UNDP, *United Nations Development Program;* UNHCR, *United Nations High Commissioner for Refugees;* UNICEF, *United Nations International Children's Emergency Fund;* WFP, *World Food Program;* WHO, *World Health Organization.*

The crucial differences between the "cluster" and the "sector" are that each area of life-saving humanitarian endeavor has a designated lead agency that would provide specialized, predictable, and accountable leadership in emergency response. Additionally, with the exception of IFRC, agencies agreed to be the provider of "last resort" where emergency gaps could not be met by humanitarian partners. Overall, the cluster lead agencies are responsible for setting humanitarian response policy and strategy in their area; developing and disseminating technical standards; analyzing response needs, gaps, and protection issues; providing the focus for overall sectorial analysis and fundraising (through the UN Appeals process); and representing/advocating for the sector. In natural disasters, the cluster lead agency must also work closely alongside government to support where the circumstances are overwhelming to national authorities and to build capacity of national counterparts (IASC, 2015).

Recognizing the importance of partnership and the enormous resources brought to humanitarian response by non-UN Agencies, the Humanitarian Reform Agenda has sought to include non-UN organizations in coordination leadership. However, despite the inclusion of IFRC, IOM, and Save the Children in cluster leadership positions, a criticism remains that humanitarian leadership remains unrepresentatively UN centric. The cluster supporting emergency response and recovery in housing is the Global Shelter Cluster, which is co-lead by the UNHCR and the IFRC. This division in leadership responsibilities means that UNHCR leads the Shelter Cluster in conflict settings, whereas IFRC convenes the cluster in natural disasters. To preserve the neutrality principle that underpins the Red Cross Movement, several terminological niceties are observed in describing IFRC's role. IFRC "convenes" rather than "leads" the Shelter Cluster and generally will not take on cluster responsibilities in an area where the Red Cross Movement is lead by the International Committee of the Red Cross—the specialist agency within the Red Cross Movement that works in conflict. In this situation, leadership could pass to UNHCR, IOM, or to the best placed NGO in-country to coordinate the response. A further clarification in IFRC-led clusters is that, following an agreement with the United Nations Office for the Coordination of Humanitarian Affairs, IFRC would not be obliged to fill the role of provider of last resort that is incumbent upon other cluster leads.

The decision about which agency coordinates the shelter response is made in-country by the Humanitarian Country Team—a group of major UN Agencies and NGOs chaired by the UN Resident Coordinator that makes initial recommendations to the Emergency Relief Coordinator and IASC about which clusters require activation based on an initial assessment of damage and in-country response capacity. The Shelter Cluster provides predictable, timely, and effective coordination services in nonrefugee situations and focuses on areas where people experience internal displacement but have not crossed an international border or have a genuine fear of persecution. UNHCR is the internationally mandated agency, under the 1951 Refugee Convention, that is responsible for multisectorial coordination in refugee contexts (UNHCR, n.d.).

While the principles underpinning shelter coordination and leadership are relatively simple, the scope of work can be enormous. Recognizing that household-level recovery begins from the immediate aftermath of the emergency, the Shelter Cluster's work includes everything related to the provision of adequate housing following emergencies. This includes the provision of nonfood items (NFIs), emergency and longer term shelter support, housing construction and reconstruction, and settlement support such as site planning and urban planning, as well as protection, disability, gender, environmental, and market and other socioeconomic issues that inform shelter and housing recovery after disasters (Global Shelter Cluster, 2016).

Following the experience of the Haiti earthquake and Pakistan floods in 2010, the IASC revisited some of the earlier concerns about effective humanitarian leadership, especially in large-scale emergencies. The resulting Transformative Agenda aimed to develop the concept of "empowered leadership," limit the longer term impacts of disasters on development, provide a common basis for assessment and response preparedness, and develop "system-wide" protocols for responding to major emergencies or "Level 3 activations" (IASC, 2016). Importantly, the common, multisector needs analyses and funding appeal based on agreed response objectives has brought a more coherent early strategy and greater international donor attention and funding to large-scale, Level 3 emergencies. Through its emergency leadership, the Shelter Cluster plays a key role in providing the strategy, objectives, needs assessment, and link with early recovery envisaged in both the Humanitarian Reform Agenda and the Transformative Agenda.

While the cluster system is, in principle, simple and flexible enough to work in both small- and large-scale disasters, the major constraints are those of funding and participation. During Level 3 emergencies, recurring concerns have been raised by national governments from the Philippines to Vanuatu that the system-wide response can itself be overwhelming and detract from the ability (and visibility) of national authorities in managing the response. Conversely, however, the designation of Level 3 activation has brought additional resources to "forgotten emergencies," such as the internal displacement crisis in the Central African Republic (ALNAP, 2016, p. 75). Smaller emergencies, however, struggle for funds as donors prioritize Level 3 activations and in the context of large-scale protracted crises, such as in Syria or South Sudan. Furthermore, funding remains driven by political considerations rather than humanitarian needs. If "forgotten emergencies" in Central African Republic, Niger, and Zimbabwe struggled for funds, the Libyan emergency appeal, which occurred at the same time, was 83% funded owing to strong geopolitical interest from the United Kingdom and the United States (Green, 2012, p. 319).

The "humanitarian system" reflects only a minority of responders most of which are Western aid agencies or affiliates. Many agencies—especially those able to raise their own revenue such as church groups—do not see the importance of collaborative, coordinated response. This is especially the case of evangelical organizations and the relief arms of political parties seeking to use crises to respond to the needs of a particular constituency or to further a political agenda. Similarly, private sector organizations are largely uncoordinated and, in countries with weak or corrupt government, are often reluctant to work too closely with national authorities or formal coordination mechanisms that exist to support government line ministries. Finally, humanitarian response may only reach a minority of those in need with processes of "self-recovery" and "resilience" little understood, especially in urban contexts where the role of cash and market mechanisms (which are well analyzed by livelihood experts) are not yet well analyzed in the context of shelter and housing.

Although there has been substantial experience of urban disasters over the last 16 years—Gujarat (India 2001), Bam (Iran, 2003), the Asian tsunami (multicountry, 2004), Pakistan (2005), Haiti (2010), Typhoon Haiyan (Philippines, 2013), Gaza (Palestine, 2014), and Nepal (2015)—this has not translated into wide-scale expertise in urban preparedness and response among traditional international humanitarian organizations. This systemic neglect of urban issues is based on common assumptions that urban recovery is often too complex, expensive, and political, and that it is best left to longer term multilateral funding institutions such as the World Bank that have closer relationships with government. For aid organizations, many of which originated in response to rural emergencies, the role of the market and access to a money economy in urban areas can also alter perceptions of vulnerability. Here, urban populations are sometimes deemed "less vulnerable" owing to comparatively higher income levels leading to aid agency perceptions that urban populations are "more resilient" and have greater capacity for "self-recovery."

The wider humanitarian system is also under strain owing to the enormous scale of refugee emergencies in the Middle East (the wider implications of the Syria Crisis), economic downturn in traditional donor nations, and relative "donor fatigue." A further problem for the humanitarian system is that it remains largely Western, and local organizations, as well as new humanitarian donors (such as China or Saudi Arabia), are not integrated into a common international system for response coordination, funding, and reconstruction. Finally, as one of the few existing coordinated international systems, whatever its shortcomings, expectations on the humanitarian system are enormous and require it to respond to conflict, natural disasters, weak governance, and long-term development challenges. These expectations are not matched by the tools, resources, or mandates of humanitarian agencies themselves, which rarely have influence beyond the provision of immediate life-saving needs. With its roots still firmly embedded in Cold War agrarian crises and conflict in sub-Saharan Africa, the humanitarian system is under strain and, like those who fund it, has shown relatively little evidence of an ability to adapt to a rapidly urbanizing world marked by longer term complex crises and climate change.

While attempts to reform the humanitarian system since the Asian tsunami of 2004 have significantly improved the quality of response leadership and coordination, the rapidly changing complexion of crises, urbanization, geopolitical manipulation, the emergence of cash transfer programming (CTP), and the desire of emerging economies to be seen as donor rather than beneficiary nations strain a system that was never intended to go beyond the provision of short-term, life-saving relief. Given the diversity of emergency situations and needs and the equal diversity of ways of responding (from individual action to NGOs to the private sector), it may be misleading to speak of a "humanitarian system" at all. Instead, as a recent study points out, on a "spectrum of coordination" from organizations that act with complete autonomy to those that work so closely that they "merge," most situate themselves in "communication" and "alignment." That is to say, organizations involved in response periodically talk with each other and share information where necessary but remain otherwise independent (ALNAP, 2015, pp. 16–18). Rather than speaking of a humanitarian "system," it may, in fact, be more accurate to refer to looser forms of association that are interconnected but not managed, such as "network" or "ecosystem" to understand the totality of response actors, including NGOs, private sector, local civil society, governments, military, and the affected population themselves.

RECURRING THEMES AND CHALLENGES IN COORDINATING DISASTER RESPONSE

RELATIONSHIPS WITH GOVERNMENT

In conflict, aid agencies can call upon the "humanitarian imperative" to deliver life-saving humanitarian assistance to respond to immediate needs despite government opposition or even in opposition to government action that may have contributed to a humanitarian crisis. However, in responding to natural disasters, government, whatever its capacity, remains the sovereign actor in the disaster-affected country. Shelter Cluster coordination consequently occurs alongside the appropriate government agency and ultimately under government direction. The official terminology for the Shelter Cluster in these situations is "co-lead" to the government "lead agency." Its role is to support the authorities temporarily until the crisis is over and additional management capacity is no longer required.

However, few if any countries have a "Ministry of Shelter," and the immediate task of the Shelter Coordinator is to find which government ministry best represents shelter interests. The appropriateness and capacity of the government ministry has a major impact on the effectiveness of the longer term response and the ability of shelter actors to advocate effectively at the highest levels of political decision making (i.e., cabinet). In most cases, shelter interests are divided between many different government departments, which can include social welfare, urban development, local government, trade and finance, infrastructure and public works, and the environment. Unlike, for example, the Health Cluster (co-led by WHO) that usually works with a clear counterpart in the Ministry of Health, the absence of a stable, recognized, and designated counterpart for the Shelter Cluster can fragment the disaster response. This can be a short-term advantage for the Shelter Cluster as it may facilitate advocacy and strategic decision making across a range of departments. However, the absence of a locally influential government lead agency means a lack of high-level advocacy and difficulties in handing over longer term recovery planning and coordination roles where there may be limited capacity to continue this function once the international system moves on or runs out of funds.

This can also be a challenge in terms of urban response as urban planning is frequently in the hands of local or city government, while the Shelter Cluster is usually lead by national government representatives, thus potentially limiting policy and advocacy influence at subnational levels.

ENGAGEMENT WITH LOCAL CIVIL SOCIETY ORGANIZATIONS AND THE PRIVATE SECTOR

While the cluster system is intended as an inclusive coordination forum based on principles of collaborative leadership and collective decision making, in reality it heavily favors international NGOs and UN Agencies.

In practice, in a large-scale emergency local civil society organizations, unlike their international counterparts, will be largely unfamiliar with international coordination mechanisms and funding appeals processes. There may be language barriers as well as cultural barriers to participation as the system favors tertiary-educated English speakers who are confident in their technical skills and ability to represent their organizations in public. International humanitarian actors, who are trained to be quick and assertive, frequently lack the ability to provide time, space, and support to representatives from local organizations to make meaningful contributions, despite the fact that local organizations frequently have greater access to affected people, respond first, remain active in the long term, and have in-depth understanding of local culture, languages, and politics. This systemic bias is reflected in global funding outcomes with a mere 1.6% of global humanitarian funding going to local actors (IFRC, 2015). An exception here is where local organizations are partnered with international ones and supported with greater resources and expertise than would otherwise be the case. As meaningful engagement of local NGOs is a key challenge for humanitarian leadership, so is engagement with the private sector.

Just as NGOs are extremely diverse in their size, capacity, and areas of specialization, private sector actors are equally nebulous. They range from high-profile corporate social philanthropy, to ease overseas remittances in times of crisis, to local chambers of commerce or shopkeepers providing relief items through market mechanisms. Crucial questions around the role of private sector organizations concern the quality and consistency of relief items provided in emergencies as well as the interaction of the international aid system with markets as humanitarian response increasingly comes to depend on CTP (Shelter Cluster Nepal, 2015a, p. 26).

PROTECTION, GENDER, AND DISABILITY INCLUSION

Shelter is frequently described as the physical embodiment of protection, and technical guidance on shelter programming often focuses on principles of social vulnerability and social inclusion rather than purely planning, architectural or engineering information. One of the ironies of the cluster system is that protection itself is viewed as a separate specialization, with its own cluster led by UNICEF, that forms a separate response strategy rather than something that is integrated more closely with other sectors.

Protection in the context of shelter considers both individual rights, including a "right to housing," and broader social themes that impact the ability of a household to recover, as vulnerability in disasters is often the product of preexisting conditions (Barber, 2008, pp. 36–37). This may relate to class, socioeconomic status, ethnicity, sexuality, disability, and questions of "locational disadvantage" such as access to markets, jobs, and financial institutions (Shelter Cluster Nepal, 2015b). Specifically, poor building materials and weak construction practices mean houses are more prone to collapse. Geographic location and the absence of political representation in capital cities or at the level of national politics can mean resources do not reach the most vulnerable or disenfranchised communities. Further, "resilience" and "self-recovery" are often dependent on social networks and connections. Those with fewer family, social, economic, or political connections—such as the elderly, people living with disabilities, or marginalized communities—are less likely to recover quickly.

Gender considerations inform all aspects of shelter programming and response and refers to the different socioeconomic roles often played by both men and women, some of which are contingent while others are embedded within cultural traditions. Different socioeconomic roles played by men and women influence access to power, resources, and the ability of the household to recover. As men and women are not uniform groups it is necessary to "disaggregate between different groups of women and men in terms of their diversity, their needs and vulnerabilities in a crisis, remembering that gender not only changes over time, but disasters and conflict can be triggers for changes in the roles and responsibilities that women, girls, boys and men have (i.e. gender roles)" (Global Shelter Cluster, 2013).

In urban response, this is further complicated by rapid processes of social and economic change, including gender roles, brought about by migration to the city. A major additional protection concern is around human trafficking and exploitation of young men and women, as well as children, for sex work and slave labor. Linking shelter, livelihoods, and a close understanding of local protection issues can promote recovery across the affected population (not just further support those who are already best placed to rebuild) as well as contribute to the prevention of human rights abuses and exploitation that can accompany disasters; however, intercluster coordination remains a weakness of the international system.

HOUSING, LAND, AND PROPERTY

Security of land tenure underpins every shelter response, is both an immediate and longer term basis for Shelter Cluster advocacy, supports timely recovery, enables longer term investment in disaster risk education, supports gender equity (as women and children are least likely to have formal land rights), and can contribute to the construction of peaceful communities by removing land ownership as a key source of communal tension. Disasters tend to exacerbate already existing social inequalities, and legal and regulatory barriers to land access and tenure can affect a household or community's ability to recover from disaster. In many countries systems for managing land ownership and cadastral surveys are weak, limited, nonexistent, or subject to traditional jurisdiction (such as *wantok* systems of collective ownership in Melanesian countries in the Pacific). The availability of formal land rights in urban areas is especially difficult in the context of urban drift, the massive increase in many cities in developing countries of "informal settlements" or slums, and intense economic pressures on land values. While shelter agencies can advocate for the rights of those immediately displaced and for longer term settlement solutions that reflect sustainable access to livelihoods, continuing this advocacy requires a long-term presence, partnerships with government, local organizations, and international development institutions such as UNDP and the World Bank. As housing, land, and property (HLP) is essential, it is well beyond the influence of the humanitarian community acting alone.

Importantly, there is an increasing legal basis for the "right to housing" that underpins the advocacy role shelter agencies and the Shelter Cluster can play in addressing land tenure issues in disaster response. These include the "right to adequate housing" in the International Covenant on Economic, Social and Cultural Rights, as well as the conflict-related refugee and internally displaced person (IDP)-based Pinheiro Principles. These rights-based approaches underpin the shelter component of the Sphere standards that provides guidance for all humanitarian agencies on minimum standards and principles in humanitarian response (NRC/IFRC, 2016, pp. 6–7). The right to adequate housing is based on the right to live somewhere in security, peace, and dignity (and the right to nondiscrimination in this context).

There are seven criteria that make up the right to adequate housing. These are:

- protection against forced evictions and the arbitrary destruction and demolition of one's home;
- the right to be free from arbitrary interference with one's home, privacy, and family;
- the right to choose one's residence, to determine where to live, and to freedom of movement;
- security of tenure;
- HLP restitution;
- equal and nondiscriminatory access to adequate housing; and
- participation in housing-related decision making at the national and community levels (UNHABITAT/OCHCR, n.d., pp. 3–4).

Crucially, these criteria also define housing—rather than shelter—more comprehensively, linking it with other core human rights such as freedom of movement. This underpins a core element of shelter programming and Shelter Cluster advocacy in general, which is that shelter and housing are embedded progressively within community rehabilitation and are not a product or commodity for distribution. This provides shelter agencies with an entry point for programming, policy, and advocacy, especially in urban settings (NRC/IFRC, 2016, p. 7).

CASH AND SHELTER

Cash is increasingly becoming the modality of choice for agencies responding to crises. CTP has been in long use by food security and livelihoods sectors, which have developed significant cash expertise, and has been used by shelter actors for some time, although not as systematically and with less sector-specific guidance (Dewast, 2016, p. 5). There are clear and significant advantages to using cash as well as some risks that are often overlooked in the rush to implement cash programs in emergencies.

These advantages include providing beneficiaries with choice, flexibility, and dignity while stimulating the economic recovery through a range of CTP options, including vouchers, cash for work, cash for rent, conditional cash, restricted cash, unconditional cash, and multipurpose cash. CTP also reduces the traditional reliance on large-scale logistics and procurement that has made shelter one of the largest and most complex aid sectors. This in turn changes the aid dynamic between shelter agencies determining what beneficiaries need and beneficiaries themselves controlling the nature and scope of assistance. CTP can underpin broader Shelter Cluster objectives of supporting owner-driven reconstruction.

The provision of choice is seen as particularly empowering for households and communities. Guidance on CTP often recommends targeting women for cash distribution as women's priorities in emergencies tend to reflect better the immediate humanitarian priorities of households and communities rather than more individualistic coping strategies frequently employed by men. This can lead to progress in women's status toward gender equality but, in an emergency context with changing gender roles, can also come at the risk of increased gender-based violence that has to be carefully analyzed and monitored by aid agencies. When accompanied by financial inclusion programs, such as branchless or mobile banking, CTP has the potential to make significant long-term development gains in poor, marginalized, and disaster-affected communities. However, as Oxfam guidance on CTP notes "without a strong analysis of the social relations framework of communities, such opportunities to empower marginalised groups may be lost" (CaLP, n.d.). Importantly, remote or particularly marginalized communities may be excluded from CTP or may not have sufficient market access to appropriate relief items to justify the use of cash.

In practice, however, there are remaining concerns about cash that need to be addressed systematically. Given the potential to affect inflation, market monitoring and analysis is essential, although this usually only occurs in specific sectors rather than across the humanitarian response. Equally, cash—especially general purpose cash grants—can be used on anything and breaks down sectorial divisions among shelter, livelihoods, food security, health, education, and other response areas. Consequently, intercluster coordination is a vital but often weak dimension emergency response. Finally, CTP is not a substitute for quality, and shelter agencies need robust training and monitoring mechanisms to ensure that CTP is effective in catalyzing the recovery process in shelter and housing.

ACCOUNTABILITY TO AFFECTED POPULATIONS

Accountability can be described as "the process through which an organization makes a commitment to respond to and balance the needs of stakeholders in its decision-making processes and activities, and delivers against this commitment." Or, more simply, it is the "responsible use of power" and is based on the overarching principle that humanitarian agencies work in the service of, and bear primary responsibility to, people affected by disasters and conflicts. Accountability in this context has five key components:

- Value accountability throughout the shelter cluster.
- Share information with all stakeholders.
- A feedback and complaints system is in place.

- Affected populations take a lead in making decisions, with support from organizational experts.
- Accountability is integrated throughout the project cycle (Accountability Working Group, 2013).

While agencies and clusters are increasingly effective at communication, inclusion and participation (especially at the community level) are debatable to the extent at which affected populations are able to "take a lead in making decisions" in humanitarian programming and in overall response strategy development. Further, critically, the word "accountable" is perhaps more of an expression of idealism than a reality. In practical terms, decision making in aid agencies more often rests with back donors, national governments, and senior headquarters staff, who are more likely to hold humanitarian actors "to account" rather than with affected populations themselves. It is, however, the responsibility of aid coordinators, donors, and cluster lead agencies to ensure that decision making and agenda setting are done by those in whose interest humanitarian response works.

TRANSITION TO RECOVERY

A major issue faced by all agencies in humanitarian response is how long to stay involved. In natural disasters, recovery could take 5 years or more; whereas, in conflicts or protracted emergencies, life-saving humanitarian interventions can last decades. Transition (and exit) from an emergency depends on a number of contingent factors, including agency mandate, availability of funding, government capacity, profile of the emergency, and ongoing need. These decisions are inevitably context specific.

The point at which coordination structures exit or transition has moved. Initially styled as the "Emergency" Shelter Cluster, the intention of clusters was that they were surge support for 3–6 months following a sudden onset disaster. After this, in theory, development actors would begin to resume longer term programs. In reality, however, the emergency phase rarely fits within this time frame; development actors are frequently unable to adjust programs to link in with the end of emergency intervention and provide continuity past the early recovery phase. Further, government weakness (as well as the absence of formal line ministries dealing with shelter or housing) means that "handing back" responsibility to state structures only occasionally produces continuity in planning and management. In this context, clusters have begun to last well beyond the initial emergency phase—in some cases for up to 18 months after the disaster. In particularly disaster prone countries or regions, IFRC has been able to support preparedness clusters, which means that mechanisms originally intended for emergency coordination are now established on a more or less permanent footing.

Where there is a need for longer-term coordination of housing reconstruction, the Shelter Cluster often tries to establish a separate body to take on responsibility. This works most effectively where government takes a lead role—such as the Earthquake Recovery and Reconstruction Authority following the Pakistan earthquake in 2005.

There are few global agencies with the capacity and emergency coordination experience to take on this role. UN-Habitat's intensely bureaucratic structure, disengagement from the humanitarian sector, and focus on it's own programs means that attempts to handover recovery coordination have been unsuccessful. Turf disputes between UN-Habitat, UNDP, and the World Bank are also unhelpful setting up longer term coordination platforms. The IOM is a major global shelter actor with leadership capacity but tends to view coordination as an additional activity to existing country programming. This "double hatting," however, can create the impression of bias and means that the specialized resources necessary for fully developing the recovery coordination role are not available (although this is also a shortcoming of donors who are often reluctant to fund coordination as a separate activity to programs).

A final disconnect is between the emergency cluster, the housing recovery platform (whatever form this takes), and multilateral institutions such as the World Bank. Housing recovery programs can take years to develop and fund and are often of questionable relevance to the local context. As a multilateral agency, the World Bank tends to engage with government and major international donors rather than NGOs and local actors, despite the fact that these organizations often have substantial available funding. In the Nepal earthquake response, NGOs in the shelter cluster represented committed and available funding of US$ 350 million for housing recovery (more than a third of the estimated total needed under the World Bank's Post-Disaster Needs Assessment).

Frequently, however, the reality is that the only assistance disaster-affected people receive is what's distributed during the emergency response phase. If this is to change, there needs to be greater investment by longer term development agencies and donors in recovery coordination beyond the emergency phase and a more concerted effort by multilaterals to link in with the resources and strategies developed by NGOs.

CONCLUSION

This chapter has provided an overview of the some of the roles, responsibilities, biases, and challenges facing the "humanitarian system" generally and the shelter sector in postdisaster settings more specifically. Significant reforms have been achieved to ensure disaster response is more effectively and systematically led and that affected communities themselves have greater determination over how humanitarian agencies work. The relatively recent advances in CTP provide major opportunities to ensure that people whose lives have been affected make their own decisions about relief and recovery while humanitarian agencies have become more sophisticated at analyzing gender and protection issues within their programs.

Key challenges remain, however. In an urban world, most humanitarian agencies remain embedded in traditions of rural emergency response, while organizations that do have an urban focus are often development focused and have little expertise in disaster response. If urban preparedness, response, and a comprehensive approach to

HLP issues are to be more effective, this development/humanitarian distinction needs to be bridged. Similarly, in the context of rising incomes and government capacity in many developing countries, meaningful partnerships between international organizations, local actors, and urban-focused government are essential to mitigating the impact of urban disasters. Finally, there is a need for greater understanding of recovery processes in urban environments and how humanitarian agencies can and should interact with the market and the private sector to ensure effective response and recovery.

REFERENCES

Accountability Working Group. (2013). *Accountability to affected populations*. Retrieved from *http://sheltercluster.org/sites/default/files/docs/ACCOUNTABILITY- two page summary-2013-11-11.pdf.*

ALNAP. (2015). *Exploring coordination in humanitarian clusters*. Retrieved from http://www.alnap.org/resource/20360.aspx.

ALNAP. (2016). *The state of the humanitarian system 2016*. Retrieved from http://sohs.alnap.org.

Barber, R. (2008). Protecting the right to housing in the aftermath of natural disaster: Standards in international human rights law. *International Journal of Refugee Law*, *20*(4), 432–468. http://dx.doi.org/10.1093/ijrl/een024.

CaLP. (n.d.). *Oxfam program quality standards: Cash transfer programming (CTP)*. Retrieved from http://www.cashlearning.org/downloads/oxfampqsforefsvlcash.pdf

Dewast, C. (2016). *Literature review on the use of cash in cluster* (online Global Shelter Cluster).

Global Humanitarian Partnership. (2007). *Principles of partnership*.

Global Shelter Cluster. (2013). *Shelter cluster Nepal guidance on mainstreaming gender & diversity*. Retrieved from https://www.sheltercluster.org/sites/default/files/docs/4_3_sc_gender_and_diversity_guidance_final_25_october_2013_1.pdf.

Global Shelter Cluster. (2016). *About us*. Retrieved from https://www.sheltercluster.org/working-group/about-us.

Green, D. (2012). *From poverty to power. How active citizens and effective states can change the world*. Warwickshire, UK: Practical Action Publishing Ltd. in Association With Oxfam GB.

IASC. (2015). *Reference module for cluster coordination at country level*. Retrieved from http://reliefweb.int/sites/reliefweb.int/files/resources/reference_module_for_cluster_coordination_at_country_level_july_2015.pdf.

IASC. (2016). *IASC transformative agenda*. Retrieved from https://interagencystandingcommittee.org/iasc-transformative-agenda.

IFRC. (2015). *Follow the money: Are funding patterns keeping pace with trends and evidence?* World Disaster Report. Focus on local actors; the key to humanitarian effectiveness. International Federation of Red Cross and Red Crescent Societies.

NRC/IFRC. (2016). *The importance of addressing housing, land, and property (HLP) challenges in humanitarian response*. Retrieved from https://www.nrc.no/resources/reports/the-importance-of-addressing-housing-land-and-property-hlpchallenges-in-humanitarian-response/.

OCHA. (2016). *Building a stronger, more predictable humanitarian response system.* Retrieved from http://www.ochaonline.un.org.

Save the Children. (2014). *Tsunami ten years on.* Retrieved from http://reliefweb.int/report/indonesia/tsunami-ten-years-stories-change-2004-2014-community-perceptions-indian-ocean.

Shelter Cluster Nepal. (2015a). *Private sector coordination pilot study report.* Retrieved from https://www.sheltercluster.org/sites/default/files/docs/sc_nepal_private_sector_cordination_pilot_study_report_final.pdf.

Shelter Cluster Nepal. (2015b). *Tool for shelter cluster partners on beneficiary targeting and prioritisation.* Retrieved from https://www.sheltercluster.org/sites/default/files/docs/prioritisation-final.pdf.

UNHABITAT/OCHCR. (n.d.). *Fact sheet 21. The right to adequate housing.* Geneva, Switzerland: United Nations.

UNHCR. (n.d.). *Leadership, inclusive coordination and effective delivery: Our shared duty towards the people we serve.* UNHCR.

United Nations Department of Economic and Social Affairs. (2014). *World urbanisation prospects.* Retrieved from https://esa.un.org/unpd/wup/Publications/Files/WUP2014-Highlights.pdf.

CHAPTER 5

Rebuilding or Repositioning: Lessons for Sandy, New Orleans, and Elsewhere

Edward J. Blakely[a]

University of Sydney, Sydney, NSW, Australia

INTRODUCTION

February 7, 2009 is literally and figuratively seared into many Australian minds with graphic descriptions of conflagrations in the Kinglake hills and many other areas outside Melbourne that destroyed 2100 homes, killed 173 people, and came within hours of racing down the hills to encroach upon the City of Melbourne. When the embers were doused, then Premier John Brumby said "They [the dead] unite us all in the task of rebuilding. Because we will rebuild [sic]" (Blakely, 2009). He was, unfortunately, wrong. *We can never and should never rebuild what was because the old settlement geographic, economic, and societal position is no longer sustainable*.

Since Black Saturday there have been fires and floods in Australia of similar impact magnitude such as the Brisbane floods in 2010 that caused damages over $1 billion killing 38 people along with subsequent well-known devastation in the Australia–New Zealand region, such as the Christchurch earthquakes of 2011, destroying one of the New Zealand's most beautiful cities killing 185 people, some of whom were immigrant students caught in a freak accident.

Finally, but by no means complete is the tragedy that hits the American Eastern Seaboard in 2012 doing more than $33 billion in damage that threatened Wall Street, the global trading center, with total closure. All these separate events have a common thread. They are in or near wealthy dense and economically important global centers. As a result, re-creating or rebuilding these places was never in doubt.

No matter how important these places are, serious questions regarding their rebuilding have to be asked, focused not on whether, but *how*, to rebuild. After a major urban disaster, we cannot, no matter how strong the popular or political rhetoric, go

[a] Known as the Recovery Czar, Ed has worked on major disaster recoveries in Oakland, California, twice 1989 and 1991 as well as 9–11 from 2001 to 2003 and recovery director in New Orleans from 2007 to 2009 and is currently honorary professor at the University of Sydney, Australia.

Urban Planning for Disaster Recovery. http://dx.doi.org/10.1016/B978-0-12-804276-2.00005-0

back to the past for superficial demographics and economic reasons. Urban systems, over the past 100 years, have sought to thwart nature and its associated processes. Urban planners and engineers facilitated the development of a regional land use pattern sprawling across the landscape. In these halcyon days, building *against* nature became a central tenant. Beginning in the 1920s and accelerating after the Second World War, urban nature-defying systems emerged, with infrastructure projects taking the form of bridges, channelization of water courses, levies, sea walls, and tunnels (to name a few) we use today. We now inhabit cities that are based upon seeking to on degrading, often wrongfully conceived infrastructure. Cities directly threatened by seas and rivers erected skyscrapers on landfill and erected sea walls and other treatments. At the same time, our regional demographic profiles are changing into an aging urban population in many developed nations, in parallel with many developing nations experiencing huge population growth and shifts to urban areas. Overall this has lead to disproportionate numbers living in fragile areas that cannot sustain the increasingly severe weather regimes brought on by even small climatic changes.

Climate change, such as those evidenced by mountain glacier retreat and other indicators around the world, is devastating for agriculture and human settlement. As humans we resist the changes in our settlements that might help our current generation survive. And if we do not make the sacrifices required to adapt and improve current settlement patterns, there stands the very real risk of condemning future generations to having to live in unsustainable ways in increasingly fragile environments. Thus, catastrophes are compounded by our current building locations, spatial planning systems, and supply and transportation chains that date back to the post–World War II era. Post–World War II building assumed a different form of settlement pattern with low-density growth in extended suburbs, or in developing nations and Europe, it failed to meet the growing autooriented cultural demands. Volcanoes and earthquakes impacting upon settlements in Europe, North and South America, and most recently the Asian Pacific, "Due to their far-reaching effects on climate, food security, transportation, and supply chains, these events have the potential to trigger global disaster and catastrophe" (Gray, 2015).

Asia is particularly vulnerable to new earthquake and flooding because of the poor design of much older and some new infrastructure that disregards basic climatic change. China's recent and continuing foods and buildings collapse in the wake of rains and severe weather and compounded by building massive dams and other infrastructure over sensitive habitat. Moreover, even well-planned and relatively dense European models adopted in other locations are now over seven decades old and require enormous investments to match rising urbanization levels and new lifestyles, combined with aging populations in central city areas. So the usual mantra from politicians to rebuild postdisaster is patently incorrect and dangerous. *We have to reposition not merely rebuild* (Lai, 2011).

WHY REPOSITION VERSUS REBUILD

My experience in disaster recovery, as an expert charged with the task of rebuilding in Oakland, California, New York, and New Orleans (Rebuild by Design, n.d.), is that the "need" to get back to normal overwhelms the opportunity to move to a smarter future. Here are several issues worthy of consideration as we embark on rebuilding cities and regions around the world.

We have an increasingly diverse population that requires houses, jobs, and socio-economic equity and security. So, the issues are joined. Continuing major disasters presents an opportunity for the region to confront the need to reposition our regions to be genuinely resilient to meet the needs of many scenario futures and not simply replicating an unsustainable past.

Leading with Information: The impulse to get back to the "old days" is strong. As a result, important information about the community vulnerabilities and possibilities is not examined before political and community forces push to re-create a version of conditions that existed before the disaster event. This nostalgia is an important social response. But we have to present the community with information on who is living in the communities and region and what they are facing. Many communities' demographics make single home rebuilding difficult in contemporary circumstances. In Japan's horrific tsunami in 2011, the vast majority of residents were senior citizens in their 1970s and 1980s (Fig. 5.1), so what is to be rebuilt for them, individual or collective dwellings?

Residential populations over 65 are often reluctant to go through rebuilding processes that can take 2 or 3 years. Further, the new building standards often mean that costs exceed homeowners' insurance. Moreover, evacuations in new events are extremely difficult for aging populations. In some cases, these seniors elect to move to new areas closer to relatives or health facilities. As a result, rebuilding often occurs with large numbers of vacant areas in blocks, making provision of services difficult and the leaving neighborhoods looking forlorn and incomplete for long periods.

FIGURE 5.1 Disaster Proof Housing San Francisco.

Author Photo: Multi-Family Podium Housing for Seniors and Modest Incomes in San Francisco above Flood level 2009, San Francisco North Beach.

Finally, some homes or apartments are simply situated in areas or built in ways that increase flooding and run off or cannot be protected from sea surge of high wind events. Relocating these units is painful, but it is necessary to protect other areas and residents. However, we are often held back by:

- **1:100 year Syndrome**—the belief that large-scale events occur only once in every 100 years; so why should we worry? If we just had an event, we will not have another one for 100 years, right? No, that is wrong. Here, 1 in 100 means a 1 in 100 chance that the event will occur in a given year. These data are revised periodically but are often not well understood by planners or residents. Cumulative impacts are particularly difficult when a zone is designated as a 1 in 100 zone without cognizance of the adjacent building that can be subsequently built well away from the planned suburb that will increase stream flows or cover over land masses that previously held or carried water.
  Moreover, many urban areas have networks of streams and waterways underground. Some of these were merely filled in like the waterways that flow under New Orleans. Several hundred years ago, what is now New Orleans was a network of islands. Over the years these islands were merged by ongoing land filling or creating a network of conduits to move water away from buildings. Both New York City and New Orleans Canal Street connote actual waterways built over a long ago. We are now at the point where rebuilding requires the resurrection or at least rethinking of how these natural systems should be used to prevent future damage and in some cases to actually improve the character of neighborhoods (Fig. 5.2).
- **Social and Economic Equity Issues**—are magnified after disaster. Communities that have few resources may be pitted against those that appear to have more resources. In many cases, this is more perception than reality. Nonetheless, some communities poorly located prestorm house more than their share of lower income, elderly or minorities. Early planning for recovery has to be sensitive to these issues. In some cases, these socially sensitive areas are the most vulnerable and relocations or other options have to be considered and handled carefully (Fig. 5.3).
  The Ninth Ward in New Orleans is among the most disadvantaged neighborhoods in the city. It is also an area extremely vulnerable to flooding as much of the area is close to many of the levees. The cost benefits of rebuilding some of the Ninth Ward communities made it hard to justify reconstruction. But the emotional issues are substantial and have real cost to families' mental and physical health as well as their long-term financial resources. Even if compensated for a lost home, a new place to live in a community with few affordable areas is difficult—not to mention the burden of reestablishing in a new neighborhood. As a recovery director, my team and I, as shown in Fig. 5.4, talked to locals working hard to find the best solution to this thorny problem. Several approaches were used in this case that can be used postdisaster in many places.

FIGURE 5.2 New Orleans Approach to Water Course Restoration.

Dutch Dialogue Public Presentation shared slides public property, New Orleans 2008.

- *Insourcing*—Rather than outsource work to large companies headquartered outside the city or even the country, we developed a process called insourcing or using local human and physical capital wherever possible. We created new rules for local neighborhood restoration by using local as the primary source of labor to restore local facilities in their community. We were creative in using community-based nonprofits as a vehicle to contract through to meet government accountability rules. This approach restored community pride as well as leading to many local innovations in replacing local infrastructure such as street lights, using nonpotable flood water for cleaning and other outcomes. Moreover, this process infused money and jobs back into the communities that desperately needed work.

FIGURE 5.3 Consultation and Problem Solving in the Ninth Ward.

Author Photos: Working with the Ninth Ward, New Orleans, 2007.

FIGURE 5.4 Kobe Local Emergency Preparedness.

Author at Kobe Safety and Survival Center, 2010, Kobe, Japan.

   - *New Neighborhoods*—Since old neighborhoods in the low-lying areas were destroyed, we had to rethink what a resilient community would look like. We commissioned local planners and architects to rethink and design more resilient communities incorporating work centers, such as community cooperative stores and outdoor markets. We also realigned houses by allowing the reconfiguration of lot lines, so larger safer homes could be assembled on higher ground in certain areas. This process was complex, but we were able to craft local rules to permit this. This approach allowed many flood victims to return home earlier and accommodate other relatives in expanded homes in their old neighborhoods.
   - *Re-creating Neighborhoods*—Since there was considerable abandoned property both prehurricane and posthurricane, we decide to leverage these assets by aggregating or moving abandoned homes onto new lots to generate more local density in communities, thus generating a large enough resident population to reach the thresholds to reopen stores and offer other civic amenities, such as libraries and reopen schools.
- **Recovery Amnesia**—sets in within the first 6 months. As the weather improves and normal life returns, the storm seems distant and the concern of residents recedes. More pressing local issues loom larger for residents, such as local elections and the like. Not only does the storm become less memorable, memories of it become increasingly distorted to justify all kinds of actions or inactions. One or two years after the event many people are in denial, so it is important to do as much as possible to change the frame of reference to long-term rebuilding as early as possible and act on delivery of cornerstone projects in place early. People want to go back home, so they place their memories of the past ahead of current realities. Once, most residents and businesses restart the events of the recent past fade and they rationalize that "we had the worst hit us, so it will be a long time before it will happen again."

There are many ways to deal with this issue. The best is a strong continuing public education program such as the San Francisco Bay area. In the Bay area annual drills and exercises are held in cities to prepare and remind residents of their duties in case of disaster. Part of these programs are graphic reminders of past disasters with local failures in the response system. Japan has a national program, which incorporates local drills in the use of evacuation routes and shelters.

Finally, school programs are important because children born postdisaster have no recall. Again, the State of California and the Netherlands are very advanced in school-based resilience education as well as disaster response training for school age children. These efforts to prepare are just as useful in repositioning as other mitigations. Memorial buildings and statues are important, but they are static so that they do not carry continuing reminders that harness public consciousness.

REPAIR VERSUS REPOSITION

Quick repairs to restore power and essential services are important, but they should not undermine the fact that these services failed. So, very early on new alternatives have to be raised in the public debate on how many public resources and some private ones might be repositioned in profound and effective ways. This means that new funding will have to be found, or other incentives such as tax breaks need to be designed to alter these vital delivery services. Other private resource deliveries, such as gasoline and food, also have to be reconfigured so that they have longer lasting supplementary power and related utilities via distributed energy, natural gas, or other forms of redundancy.

Future Proofing goes well beyond good environmental design. A future that includes a more diverse population in terms of residents' age and in parallel diverse building typologies in many neighborhoods will allow future change to occur more readily. The reasons for this are many, but the most important one is that this diversity brings younger and more able populations into communities, thus making them more stable by increasing local retailing and improving the chances of getting all ages out of harm's way in an extreme event.

Environmental design is critical. Mechanisms to facilitate neighborhoods become more self-reliant with local, decentralized water and power with insulated power producing housing, sometimes described as distributed systems (as opposed to more brittle centralized systems). Every community can and should have some food growing and food storage facilities, as well as inward evacuation systems using schools, churches, and local facilities as the first refuge for the able to sustain their inhabitants during large-scale catastrophes.

Rebuild by Design, which was created after superstorm Sandy, is a nationally financed effort challenging US communities to come up with creative solutions using natural or innovative processes to both rebuild but reposition and make communities smarter economically and more resilient to natural and man-made disasters. Rebuild by Design has captured the collective imagination of 141 cities and involved over 100 local, state, and federal government agency partnerships in coming up with collaborative processes that restore devastated communities by crafting projects that make them far safer for the present and alter the dangerous course of repeating past disasters (Rebuild by Design, n.d.). Another example is Japan, which has a network of national disaster preparedness and prevention centers that work with provincial and local governments to create innovative responses to the multitude of disasters that a great nation faces. Preparing and learning from world disasters allow Japan to preconfigure assets well ahead of disasters. Even with all of this preparedness the Fukushima Daiichi tsunami overwhelmed national capacities.

COMPETING RECOVERY VISIONS

There are competing visions of how to handle large-scale disaster events. Some proposals are entirely based on extensive environmental remediation, whereas others rely more on improved technology and engineering such as sea walls and barriers. The Dutch approach is to use both soft and hard infrastructure where appropriate and to invest in more community-level programs so people can live with water using sensitive design. Dutch schools have an extensive water curriculum to educate future generations on water as a way of life, thus reducing the postevent amnesia that affects all of us as the lessons from a catastrophe recede in public memory.

Dueling Plans—In the New Orleans case, within weeks, local groups, architects, planners, and communities commenced replanning efforts. This is a good thing. However, in many instances the communities used different information bases for their plans, thus creating enormous confusion over possible futures. While getting as many ideas as possible on the table is important, it is also important to generate them from a common base at least. A common scale and scope for plans is also essential, as some plans actually disrupt other communities, generating new conflicts. So, to the extent possible, a regional planning framework that set common information and templates is a good idea. To the extent possible, planning information and community as well as local leader education/information should precede and be part of ongoing communications to act as a base for all plans. If possible, a common planning repository should be developed so that everyone at every level can keep track of the planning process (Fig. 5.5).

In New Orleans, we were faced with multiple neighborhood and regional plans. All these plans had good ideas about the needs of the past and restoration

FIGURE 5.5 Target Area Zones Recovery Map, New Orleans.

Courtesy New Orleans Office of Recovery Management, 2008.

of places and communities, but few contained a holistic view of the entire city and region combining social, economic, and environmental visions. Eventually, we were able to fashion an overall plan incorporating aspirations with sound economic and environmental knowledge with considerable community input (Blakely, 2012).

ECONOMIC REPOSITIONING IN RECOVERY

Economic revitalization and repositioning are essential. A megastorm not only disrupts the current economy, but some firms also do not survive financially and physically. In many cases, small community-based and even regional firms lose too much of their customer base to continue. In addition, the world will move on of its own accord while communities recover. During national disasters and the associated long recovery periods, the central firms in the local economy may lose their competitive edge. For example, Kobe, Japan was one of the top five shipping cities in the world when it suffered a disastrous earthquake in 1995. By the time the city reached normality almost a decade later, it was no longer an important shipping center and many of the auto firms dependent on Kobe's port had relocated to other port cities. Two decades after the earthquake, the demographics reflect the manner in which change has occurred: only 40% of the residents in the city were living in the city at the time of the earthquake and few residents work in any prequake firms. Kobe, like many places, had to find a new economic base while it rebuilt. Another illustration is Aceh, Indonesia, where a Tsunami wiped out local fishing villages in 2004. Experts in the national recovery agency and international experts determined that the safest course of action was to build new villages a kilometer or so from the ravages of another tsunami. For the local villagers, this was untenable economically. Village fishermen needed to be close to the sea to insure catching their daily income. So, they moved as close to the ocean as they could. They knew the danger, but they calculated mentally and from old stories another tsunami would not hit for many years.

After superstorm Sandy in New York, there were fears that the New York Stock Exchange was located too close to the Harbor. While the Exchange has not moved its historic building, many of the functions and employees are now in far-flung locations across the Hudson in New Jersey and others as far away as Dallas, Texas.

REPOSITIONING OPPORTUNITIES

Disaster recovery means balancing the past with the future. It is important to show a new path to a better future, rather than suggesting options that fail to recognize preservation of the past as an important element of finding improved and more resilient communities. The way to deal with this is to create a new long-term direction to reconfiguring the rebuilt assets to confront new economic realities. Some sectors will have to rethink where they are going and others will be a position to add new value to the economy. Health care and construction can use large events to develop better and smarter delivery systems reaching new populations, or with new products and services. New firms can arise in the emergency services fields with export capacity as well as new environmentally sensitive products and services, including food production and distribution. Transportation infrastructure along with telecommunications is a key sector for renewing with potential for new spin-offs. New Orleans is an apt illustration. During the recovery, a group of community leaders led a campaign to use the rebuilding of the city central district into a new modern biomedical center. New Orleans is now home to one of the US largest health and biomedical complexes. In this case the disaster created a new opportunity that the community seized—not without opposition and struggles—to transform the city's base economy away from tourism and shipping, which generated more well-paying jobs and revitalized the heart of the city (Fig. 5.6).

Underlying issues can often hamper recovery. Many communities have long-term issues with deep roots that need serious examination postdisaster. One of the most difficult and vexing ones can be the organization of local government(s). Local governance in the United States, for example, is quite fragmented, slowing responsiveness and limiting effectiveness both during and after a disaster. While issues of government may seem too hard, these issues need to be tackled at the time when the problems are most visible. Similarly, the apparent hydraheaded monster of utilities accountability and leadership needs to be addressed while it is fresh in the recovery process.

Several other recovery repositioning programs in developing nations are worth noting. Chile is a prime example of using postearthquake opportunities to upgrade slums and generate new housing finance and funding strategies that generated new construction methods that were quicker and more resilient than the former housing approaches. The Chilean model has gained international recognition and is now

FIGURE 5.6 New Orleans Downtown Hospital Complex.

Courtesy Veteran Administration Architects non-copyright public document.

promoted by the World Bank and donor agencies as the best approach to rebuilding the entire community and not rebuilding with the same social and economic inequities along with continuing one house at a time on the rebuilding approach. We see too frequently, in Chile over 12 months, a program was developed to create a series of temporary shelter villages, and a system of recovery housing subsidies were established; risk-based land use plans were conducted in various coastal areas; a finance plan was adopted; changes to the national emergency management agency were made; and rapid payment of insurance claims were completed (Siembieda, Johnson, & Franco, 2012).

While national governments are often highly engaged as Chile demonstrates, recovery is not free of the challenges associated with receiving national government money. Local agencies will usually have to form new structures such as public private partnerships to deliver rebuilding projects. Much of the long-term financing will require state and local as well as private matches. Moreover, issues from sea walls and some environmental mitigation will attract federal resources, but much of the local rebuilding will need local or state funding over a long and sustained period. California's building retrofit programs post–Loma Prieda Earthquake of 1986 required substantial building buttressing in public and private buildings, including incentives to improve home stabilization systems. California issued a large voter-backed bond financing scheme to achieve this informing citizens that it might take more than a decade to complete. Creative thinking in an antitax environment such as the United States is required. California is an illustration in which the public is prepared to fund items such as storm surge property removals, coastal and waterway restoration, and evacuation routes and home physical design security that insulate them from known dangers with visible, measurable progress and accountability as part of the package.

LESSONS

Based on direct lessons learned, this chapter aims to challenge the paradigm of "rebuild right now here in the same way things stood in the past." National legislation in many nations requires or strongly encourages rebuilding the same public buildings where they were, to serve the same mission they did in the past. In fact, the past is not prologue. In too many cases what was built years ago does not serve a new changed world. Moreover, older style building, which was led by engineers, has now often demonstrated itself to be environmentally destructive, so repeating the conditions that contributed to disaster is not only lacking sense, but it is also the precursor to new disasters based on a changed ecological terrain and climate change.

So, the intelligent thing to do postdisaster is to look to the future and mitigate environment, social, and community political issues that harm the area from moving forward. As described earlier, physical disasters frequently reveal cleavages in the social order that need to be dealt with so the community is truly restored to a more robust functioning and health. In addition, building on disturbed earth systems will often lead to future hazardous events. Water systems will often find ways of doing damage unless new ways are found to live with and to accommodate it, rather than to try to conquer them above or below ground.

Repositioning we suggest has four elements:

1. Making peace with nature—Since political geographies do not define natural systems, a deep analysis of the geotechnical environment is necessary. The rebuilding strategy has to take the entire natural system into consideration before any structures are put in place or we are simply creating the path of destruction for the next disaster.
2. Repairing social structures first—Socioeconomic disparities are laid bare post-disaster. These cleavages in the social system will impede and perhaps destroy the rebuilding process unless they are dealt with head on. This means the people deeply affected by the tragedy must be directly involved in developing the options for the future. And where possible and feasible, local people should be employed in the rebuilding process.
3. Economic rebalancing—The local economy losses its competitive position after any disaster. Money moves away from damager. New trade routes are established and firms move away. So it is important not only to restore the old economy but also to find a new future economy. Repositioning the local assets in new way is critical to this process as we have detailed here.
4. Collaboration and competition—It is critical for damaged communities to regain their competitive position. When they are damaged, it is too easy to become a mendicant. There is not future in feeling sorry for the people or the place. Thus, new partnerships and collaborations have to be formed to move the community back into the main stream of regional and national and perhaps international participation. The world does not stop for any place.

FROM HERE TO WHERE

The case to reposition is clear. But before deciding to reposition, a region and its communities must have a firm grip on where they are with respect to their current geophysical system having assessed what needs to be changed for a more sustainable future. Similarly, regional economic and social accounts need to be undertaken to explore what options and opportunities exist for transforming the region in the future along with what strategies are required to get to a new economic regime. Thus, if the region (cities sharing a common geoeconomic shed) is already operating as a system, it has the capacity to implement needed changes postdisaster or incrementally according to some form of regional planning process. In essence, long-term well-informed plans are the bedrock of constructing and repositioned future after any form of calamity.

Planning for the future is the best antidote postdisaster. Plans are the glue that holds the community together and the springboard for constructive action. New York's *PlaNYC* was invoked after superstorm Sandy calming city residents and placing city leadership in a unique position to deal with longstanding capital infrastructure, economic, and social policy needs (The City of New York, 2016). Cities, regions, and communities that plan for a future can move quickly to rebound from disasters. Those that fail to design a future usually languish postdisaster for long periods because they have no common platform to deal with the present tragedy and run the risk of trying to get to the future by looking for the past.

Resilience is the new concept, which incorporates the ideas articulated here. It is a more useful and powerful construct than sustainable or smart because it deals not only with mitigation but also with the construction of new ways to build for the present and recover in the future.

REFERENCES

Blakely, E. (2009). *Life after Black Saturday. The Sydney Morning Herald.* Retrieved from http://www.smh.com.au/federal-politics/life-after-black-saturday-20090225-8hv0.html.

Blakely, E. (2012). *My Storm: Managing the recovery of New Orleans in the wake of Katrina.* University of Pennsylvania Press.

Gray, R. (2015). *Will a volcanic eruption destroy humanity? Scientists warn that world must begin preparing for explosive global catastrophe.* Retrieved from http://www.dailymail.co.uk/sciencetech/article-3039652/Will-volcanic-eruption-destroy-humanity-Scientists-warn-world-begin-preparing-explosive-global-catastrophe.html#ixzz4HSbUDmxQ.

Lai, E. C.-Y. (2011). *Climate change impacts on China's environment: Biophysical impacts.* China Environment Forum: Wilson Center.

Rebuild by Design. (n.d.). *Rebuild by design.* Retrieved from http://www.rebuildbydesign.org/what-is-rebuild-by-design/.

Siembieda, W., Johnson, L., & Franco, G. (2012). Rebuild fast but rebuild better: Chile's initial recovery following the 27 February 2010 earthquake and tsunami. *Earthquake Spectra, 28*(S1).

The City of New York. (2016). *One NYC.* Retrieved from http://www1.nyc.gov/html/onenyc/index.html#oneNYC.

CHAPTER

6 Finance, Insurance, and Facilitation of Recovery: Should the Role and Responsibility Assigned to Government Be to Assert Control Over Long-Term Planning?

Sarah-Alice Miles
Amersfoort, Netherlands

INTRODUCTION

It is well understood that the goals of disaster management aim to reduce or avoid losses from national and local hazards, to assure prompt assistance to victims, and to achieve rapid and effective recovery.[1] Recovery involves rebuilding the community and society, such as repairing or reconstructing housing, property issues, employment, and restoring essential infrastructure so as to return a system to a functional and less vulnerable state (Blaikie, Cannon, Davis, & Wisner, 1994; Coppola, 1999). Berke, Kartez, and Wenger (1993) stated that of these four phases, recovery is the least understood and perhaps the most challenging stage of the disaster cycle. An important aspect of effective recovery is taking the advantage of the "window of opportunity" for the implementation of mitigative measures that might otherwise be unpopular (Alexander, 2002, 2008). The effectiveness of response and recovery is eroded if performed in the absence of a comprehensive regime of preparedness and mitigation (Coppola, 2006). While there is a considerable amount of literature with significant focus on predisaster preparedness, postdisaster recovery and reconstruction have not received the same emphasis. Poorly coordinated and planned recovery can lead to long-term victimization and abandonment of the affected population. Recovery requires clear allocation of responsibilities, defined mandates, and

[1] Disaster recovery begins with stabilization and ends when a community has reestablished normal social, economic, and political routines. Disaster recovery encompasses multiple activities, some of which are implemented sequentially and others are implemented simultaneously. Attempts to define differentiated phases of disaster recovery are inherently limited in their validity, so researchers have generally been less concerned about time phases (e.g., short-term recovery vs long-term recovery) than about the specific recovery functions that must be performed.

Urban Planning for Disaster Recovery. http://dx.doi.org/10.1016/B978-0-12-804276-2.00006-2

the development of regulatory frameworks calling for close collaboration between professionals, agencies, and interest groups from a diverse range of disciplines and perspectives. Without smooth collaboration between them, recovery is likely to be slow and painful. Not long before the first Canterbury earthquake in 2010, Rotimi outlined a series of deficiencies, declaring that the statutory basis for coordination was inadequate with respect to the recovery legislation and supporting frameworks. Long-term recovery was not supported by legislative powers and consequently local authorities were to take the lead in recovery.

It is often the case that local government has responsibility for rebuilding public facilities and infrastructure postdisaster, while the private sector is generally responsible for rebuilding houses and businesses. It is the private sector, with the assistance of government, which usually is seen to have responsibility to restore overall economic vitality. The big issue remains how to establish a sensible way of planning for the recovery, postevent, both economically and politically. This is complex. The challenge for a nation is to work together regardless of the many differing agendas of various stakeholders and legislators, within both the private and the public sectors (Miles, 2016).

In this chapter the author explores the role of central government in the recovery process and to what extent it has taken control with respect to planning postdisaster after the recent Canterbury earthquakes in the New Zealand (2010–12). She concludes that the outcome of the recovery process has been less about the products of violent seismic events themselves than the result of political decision-making and the questionable actions of an "absent" government. She poses the question, how broad should the role and responsibility assigned to the government be, with respect to the control of long-term planning, postdisaster? She argues that the New Zealand central government failed in its responsibility to keep checking the growing inequalities, and it should have better protected its earthquake victims from corporate abuse and asset loss. The author examines requirement of the growing reliance on public–private ties from the central government if disaster management is to function well.

SITUATIONAL HISTORY

New Zealand experienced a series of devastating earthquakes between 2010 and 2012.[2] At 4.35 a.m., on Saturday, September 4, 2010, the first of a series of major earthquakes struck the Canterbury region of the New Zealand's South Island and measured 7.1 on the Richter scale. Within seconds the Canterbury region was brought to a stark new reality for which it was woefully unprepared. On Boxing Day 2010, more than 24 earthquakes, including another large 4.9 quake, shook the city. Then, the *coup de gràce*—a second major, shallow earthquake struck, at 12.51 p.m., on February 22, 2011, measuring 6.3 on the Richter scale. It was said to be among the 10 strongest earthquakes recorded in the New Zealand. Extensive damage was caused,

[2] See http://www.christchurchquakemap.co.nz/all.

crippling Christchurch City and its suburbs.[3] It struck close to the Central Business District, the heart of the city, in the middle of the day when people were at work. Many buildings, already weakened by the previous quakes, collapsed. Liquefaction[4] immediately posed huge challenges for access and remedial work. This event cost 185 people their lives, with thousands more injured.

As of 2017 the Canterbury earthquake recovery is still ongoing. The main stakeholders involved in the Christchurch recovery are the government, the Earthquake Commission (EQC),[5] Christchurch Earthquake Recovery Authority (CERA),[6] Regenerate Christchurch (replacing CERA in April 2016), the Christchurch City Council (CCC), the insurers/reinsurers, the construction industry, businesses, and the affected population. The Civil Defence and Emergency Management (CDEM) Act 2002 and the National CDEM Strategy detail the management of hazards in the New Zealand. Accordingly, when a natural disaster affects a community in the New Zealand, in the first instance it is the Local City Council that is responsible for providing comprehensive and integrated emergency management.[7] This involves preparedness, mitigation, response, and recovery. Christchurch City, before the earthquakes, had a strong mayor–council form of local government.

After the first earthquake, on September 4, 2010, recovery-related activities were undertaken according to plan. Soon after the earthquake, a state of "local emergency" was declared within each of the three affected districts (Selwyn District, Waimakariri District, and Christchurch City), and persisted until September 16, 2010. Two days later, Minister Gerry Brownlee was appointed as the minister responsible for the Canterbury Earthquake Recovery and tasked with "trouble shooting." An ad hoc

[3] Christchurch is one of the New Zealand's largest cities with a population of approximately 340,000 people. The City Council consists of 13 councilors elected from seven wards and is presided over by the Mayor who is elected at large.

[4] Liquefaction occurs when loosely packed; water-logged sediments at or near the ground surface lose their strength in response to strong ground shaking. Liquefaction occurring beneath buildings and other structures can cause major damage during earthquakes.

[5] The present EQC is a Crown Entity as defined by the Crown Entities Act 2004 and is governed by a board of eight commissioners who are responsible to the Canterbury Earthquake Recovery Minister. Just before the Canterbury earthquakes EQC had 22 permanent staff, all based at the Commission's office in Wellington. The Commission's primary objectives are to administer insurance against natural disaster damage as provided for under the EQC Act 1993, to facilitate research and educate the nation about matters relevant to natural disaster damage, to manage the Natural Disaster Fund, and arrange reinsurance. It designates itself as a public benefit entity and constitutes one of the Crown Financial Institutions that manage large funds at arm's length from government—in this case the Natural Disaster Fund. Any deficiency in the Natural Disaster Fund to meet the liabilities of the commission is to be made up by the government by way of a grant or advance. The viability of the Natural Disaster Fund is critical for the effectiveness of EQC.

[6] Economic Recovery Program for Greater Christchurch: A foundation for economic recovery and growth in greater Christchurch. Christchurch: Canterbury Earthquake Recovery Authority (CERA).

[7] The Ministry of Civil Defence and Emergency Management (MCDEM) is responsible for disaster response and recovery at a national level but planning for an implementation of disaster response is through CDEM groups at the local level. Local authorities and their communities lead response and recovery. The CDEM act only addresses recovery during the state of emergency phase.

Cabinet Committee was established, and a process was initiated to develop legislation to assist in the management of aspects of response and recovery. Parliament enacted the Canterbury Earthquake Response and Recovery Act 2010 (CERRA) on September 14, 2010, assented to under extreme urgency. CERRA was rushed through Parliament on the basis that it was required to hasten a speedy recovery. Thus the government had "moved in" quickly and the population believed that assistance was well on its way.[8] The CERA was also created under CERRA, the week before the second major earthquake, in February 2011. CERA was effectively a government department reporting directly to the Earthquake Recovery Minister.[9] Initially the CERRA Bill received virtually unanimous support from the parliament. However, within a fortnight a group of 27 constitutional law experts from all six New Zealand university law faculties had issued an open letter detailing concerns about the breadth of the powers granted under CERRA, calling it a "dangerous precedent" as it abandoned established constitutional values and principles.[10] The danger was said to lie in a poorly framed general power to regulate, with weak legislative oversight and limitations on the review of that power by the courts. It was later repealed by the Canterbury Earthquake Recovery Bill 286-1, 2011, which in turn, put in place certain checks to guard against the inappropriate use of the powers given to the minister and CERA.

On February 23, 2011, a "national state of emergency" was declared. The New Zealand Prime Minister (John Phillip Key) addressed media stating that he had faith in the leadership in Christchurch, but national emergency status would "give more control to the government," enabling the government to direct local, national, and international resources to achieve the "best possible response in the shortest timeframe."[11] A national state of emergency provided the authority to suspend some normal functions of the executive, legislative, and judicial powers, and to order government agencies to implement emergency and rapid response plans. It was said that as soon as the civil defence emergency period ended, this authority would be exercised working in close support of and in cooperation with the Mayor of Christchurch and the Christchurch Civil Defence team. Earthquake response and recovery functions were to be transitioned out of the Emergency Operations Center. However, this did not happen and instead local government and provisions were quickly superseded

[8] http://www.stuff.co.nz/the-press/news/christchurch-earthquake-2011/canterbury-earthquake-2010/4674342/Govt-to-help-quake-homeless; http://www.stuff.co.nz/the-press/news/4096400/Earthquake-city-council-advisories.

[9] CERA was presumably set up to enable a faster mechanism for recovery. The United Nations Development Program 2006 details what that new structure should focus on, including the formulation, implementation, oversight of recovery, the monitoring of progress, and the establishment of a permanent dialog and consensus space with civil society. Opposition parties, the private sector as well as international cooperation agencies maintain transparency, accountability, and good governance. Any new structure should have as its aim the avoidance of undermining already existing institutional frameworks or well-functioning good governance mechanisms.

[10] For additional reading see Jonathan Orpin, Constitutional Aftershocks, *New Zealand Law Journal*, November 2010, pp. 386–388, and Dean Knight, Canterbury Earthquake Response and Recovery Bill: Constitutional Outrageous, Fairness and Justice, Victoria University of Wellington, September 14, 2010.

[11] www.stuff.co.nz/national/christchurch-earthquake/4693868/Key-announces-national-state-of-emergency.

by central government in the form of CERA.[12] Despite lines of responsibility having been thought to be clearly delineated, they became alarmingly confused and it was not long before the "cracks" began to appear in organizational processes.

In the Agenda Order Paper of the Canterbury Civil Defence Emergency Management Group Joint Committee (December 13, 2010), several issues arose in response to the local state of emergency. These included the coordination of information management problems at local and regional levels (which were also evident at a national level), and the manner in which the legislated and planned role of the Ministry of Civil Defence Emergency Management was to be diminished in both response and recovery. In addition, it was believed that the new legislation, the organizations, the processes and roles put in place, although intended to expedite and streamline response and recovery, actually interfered with and undermined arrangements that were previously put in place. Indeed, it seemed that the authority and the responsibilities of local government had been superseded by CERA.

CERA was given a mandate for 5 years. CERA's special powers were to cease from April 2016, when the 5-year Canterbury Earthquake Recovery Act 2010 expired. During that period CERA has been continually criticized by the affected community for having had too many powers to intervene in local recovery.[13] In the new arrangement, CERA is to be assimilated into the Department of the Prime Minister and Cabinet. The Christchurch City Mayor has publically thrown her support behind the government's new plan, but hopes it will not take another 5 years for the CCC to regain full power.[14] In July 2015 the City Council again sent the government a strong message that it no longer wanted to be a backseat passenger in its City's recovery, "It is time for a transition back to local leadership and decision-making."[15] "Regenerate Christchurch" is now the new entity to take the city into the future, and a new law will be introduced to take over the management of the central city rebuild from CERA.[16] In the recovery phase, the Mayor and City Council have attempted to take back some power,[17] and it has been agreed that a more collaborative approach is required with the inclusion of the council's development authority, Development Christchurch.

[12] CERA's task was to lead and coordinate the recovery efforts of three councils, central government departments and crown entities, infrastructure providers, business, local community, constructions firms, and Environment Canterbury including reconstruction priorities and compulsorily acquiring land, entering premises and undertaking works, and demolishing and disposing of dangerous buildings.

[13] http://www.stuff.co.nz/national/politics/78952664/Five-years-of-Cera-Success-or-failure.

[14] http://www.stuff.co.nz/the-press/business/the-rebuild/69862124/new-legislation-to-replace-canterbury-earthquake-recovery-act.

[15] http://www.stuff.co.nz/the-press/news/70503456/government-needs-to-step-back-says-christchurch-city-council.

[16] http://cera.govt.nz/news/2015/greater-christchurch-regeneration-bill-introduced-19-october-2015.

[17] http://www.stuff.co.nz/the-press/business/the-rebuild/69943847/council-claws-back-rebuild-power; http://www.stuff.co.nz/business/industries/68152533/christchurch-city-council-to-set-up-rebuild-agency.html; http://www.stuff.co.nz/the-press/news/70503456/government-needs-to-step-back-says-christchurch-city-council; www.stuff.co.nz/business/rebuilding-christchurch/6170837/Earthquake-Minister-scolds-council; www.stuff.co.nz/the-press/news/christchurch-earthquake-2011/6696503/Govt-poised-to-seize-control-says-Dalziel.

CERA and its minister, who are charged with earthquake recovery, have remained essentially unpopular throughout the post-recovery phase. CERA's command and control thinking has not sat comfortably with the population or the business community. There is also a strong sentiment among the population that the central government-managed recovery has "been left to the private markets" allowing the population to fall victim to private enterprise. Again this begs the question as what should be the central government's role postdisaster?

PERCEPTIONS OF GOVERNMENT

Immediately after the earthquakes, the New Zealand government adopted a traditional model for disaster response and then continued to progress this model throughout the recovery phase using CERA as its main vehicle. This model presupposes that government is the most reliable actor because societal chaos is likely to result after disaster and local government would have difficulty setting the priorities. The model imposes a strict organizational hierarchy and relies on "clearly defined objectives, a division of labor, a formal structure, and a set of policies and procedures to fulfill disaster and recovery operations" (Schneider, 1992). This is believed to simplify an otherwise complex process of policy-making. However, the weaknesses of this top-down model has been that it restricts the responsibilities of local government and it fails to allow people to participate in the process of recovery or to adapt to the imposed changes. The gap represented by public norms and bureaucratic norms is ultimately the key variable in determining how well the disaster is managed (Schneider, 1995).

Research has shown and it is now an established theory that a community with good social capital records the highest satisfaction rates for (new) town planning and has the speediest recovery rates. Societies with democratic governance, that is where civil society groups and non-profit organizations and other non-state actors can work both independently and in collaboration with the state/government, are those with higher potential for economic growth and better management (Brinkerhoff, 1999; Millen, 2011; Pelling, 2003; Vallance, 2011). The role of community leaders is prominent in utilizing existing resilience and social capital in the recovery process by facilitating collective decision-making (Nakagawa & Shaw, 2004). Where there is feedback from civil society, better policy develops and negative feedback from communities is useful because it forces policy makers and developers to rethink their assumptions. A failure to be heard, however, produces a sense of democratic disarmament together with a lack of transparency on part of the government, and it becomes a source of frustration, apathy, and hopelessness. There has been much feedback from Christchurch residents and the CCC about the top-down model imposed on the community in the form of CERA. These complaints did not have any impact. Instead, the focus of the government has been primarily a financial economic one, centered on restarting and facilitating business. A survey was carried out by the Christchurch Press, and the overwhelming view was that Minister of Earthquake Recovery had "taken over Christchurch."[18] In the past the CCC was, and is, seen to be playing "second fiddle" to the government's earthquake rebuild machinery. Many believe that there has been a general downgrading of the city and the regional councils. Wellington government seems unable to trust the regional councilors to come up with the outcomes; it presumably wants and has even threatened, if necessary, to sack "dysfunctional elected representatives"[19] sitting on the CCC. Without proper consultation the government decided that it would revamp/rearrange Christchurch schools,[20] introducing privatization without a mandate to do so and in addition it sought to control the building consent process[21] and priced the sale of city assets.[22] These actions have fundamentally changed the Christchurch regional power base. The minister holds day-to-day power, and the final political and executive power rests with the Prime Minister. This present power constellation is a clear indication that there is a need to restore democracy to the city. Yet, as time passes, the population becomes more resigned to its apparent inability to have a say in the recovery process. Even city councilors complain that they are left out of "the loop" in discussions of cost sharing for the city rebuild, including who will pay in the long term for a series of "anchor" projects, mooted by the government.[23]

SOCIAL COHESION

Over the years the New Zealand government has undergone a steady but perhaps little noticed transformation. Its traditional democratic processes and institutions have become marginalized and nongovernmental organizations are now more central to public policy. The consequences are that governments now assign many of the responsibilities of other models of governance, with those of private enterprise and with (non) profit organizations. These complex relationships have also caused a muddying of intergovernmental relations and made it difficult to determine who is responsible for oversight and who is actually making the decisions. In effect, over the last decades we have seen a steady "privatizing" by government. This produces an ideological misfit between obligations toward the care for the population and shareholder profit required by corporate operators. As a result, the government appears to be indifferent to citizens' needs, and elected officials fail to exert sufficient control over corporate interests. Evidenced in the introduction of charter schools,[24] the selling-off

[18] See Brownlee as clear No 1, *Christchurch Press*, May 4, 2013.

[19] http://www.radionz.co.nz/news/regional/269002/proposed-canterbury-council-structure-questioned.

[20] http://www.ppta.org.nz/index.php/-issues-in-education/charter-schools/2119-no-charter-schools-nz.

[21] http://www.stuff.co.nz/national/8864976/Emergency-meeting-last-chance-for-quake-council.

[22] http://www.scoop.co.nz/stories/AK1504/S00394/consultation-on-asset-sales-an-undemocratic-farce.htm.

[23] http://www.stuff.co.nz/the-press/news/5627660/Councillor-quits-felt-almost-irrelevant.

[24] http://www.stuff.co.nz/dominion-post/news/local-papers/the-wellingtonian/opinion/8551770/Charter-schools-disquiet-grows.

of large numbers of state houses[25] and the compulsory acquisition of land[26] were demonstrated by an apparent lack of action directed at corporate interests, even when those interests were clearly contrary to the well-being of the catastrophe-challenged population. The result has been that in reality, the expectations of a disaster-stricken population remain rooted in a past that no longer exists. It is for this reason that local governments must become key players in the recovery. Local governments are the advocates for the populations they represent. Present day capitalism cannot be simply about accumulation of wealth, it must also represent societal aspirations—aspiring to produce a better and sustainable society, a healthy, fully functioning society. Free enterprise cannot roam without a regulatory framework beyond politics. There must be a middle ground. The free markets can solve many of our problems but only if they are operated responsibly and in a humanitarian way.[27] In Christchurch today, many feel that there has been a humanitarian versus cost-control conflict and the affected residents appear to be on the losing end.[28]

In nations with small populations and strong central government, such as the New Zealand, there are often fewer decision-takers and they are placed at a higher level in the response hierarchy (Miles, 2016). They have a tendency to interfere and

---

[25] http://www.stuff.co.nz/dominion-post/comment/editorials/69809849/Editorial-Governments-state-housing-sell-off-gets-stranger;http://www.radionz.co.nz/news/political/296918/govt-gets-power-to-sell-social-housing.

[26] www.stuff.co.nz/business/rebuilding-christchurch/6831407/Compulsory-acquisition-could-reconfigure-Christchurch-CBD;http://www.interest.co.nz/bonds/66066/high-court-rules-governments-red-zone-land-buyout-offers-not-lawful-brownlee-will-appeal; http://www.listener.co.nz/current-affairs/christchurchs-game-of-zones/.

[27] Insurance firms must stop messing with our lives, Ross Williamson, Rangiora, Letters to the Editor, The Press, February 2, 2013; Clients lose plot at years of delays: Insurance anger leads to threats, as people with unsettled claims reach their wits' end, Cecile Meier, *Christchurch Press*, December 24, 2014; Repugnant insurance companies earn our distrust, Tom O'Connor, Letter to the Editor, *Christchurch Press*, January 19, 2013; CERA "using bully tactics" to get land, Martin van Beynen, *Christchurch Press*, September 7, 2014; www.stuff.co.nz/national/christchurch-earthquake/4919892/Cera-labelled-militaristic; Residents: Why are we still living like this? Blair Ensor and Georgina Stylianou, *Christchurch Press*, March 5, 2014; Housing dearth forces poor to sleep in cars, Fairfax Media, March 29, 2012: retrieved from tvnz.co.nz/national-news/housing-dearth-forces-poor-sleep-in-cars-4804904; http://www.radionz.co.nz/news/national/101884/hundreds-sleeping-in-christchurch-airport-terminal.

[28] http://www.stuff.co.nz/the-press/news/67483938/Council-warned-over-social-housing; www.stuff.co.nz/national/christchurch-earthquake/6734583/Rental-shortage-State-must-step-in; http://www.radionz.co.nz/news/regional/267010/christchurch-faces-big-rates-rise; http://www.stuff.co.nz/business/72002002/Only-1-per-cent-of-EQC-repairs-done-with-building-consent; http://www.stuff.co.nz/the-press/business/the-rebuild/72283618/extent-of-shoddy-quake-repair-work-exposed-in-reports; http://www.stuff.co.nz/the-press/news/10441327/Cheap-fixes-devaluing-thousands-of-homes; http://www.stuff.co.nz/the-press/news/christchurch-earthquake-2011/67919822/Christchurch-quake-survivors-and-the-long-road-to-mental-recovery; www.stuff.co.nz/the-press/news/christchurch-earthquake-2011/7074536/Red zone-residents-threaten-UN-action; https://thechristchurchfiasco.wordpress.com/2015/04/02/major-nz-companies-named-in-human-rights-claims/; www.stuff.co.nz/business/rebuilding-christchurch/7683139/Laywers-called-in-over-red zone-offers; www.stuff.co.nz/the-press/news/christchurch-earthquake-2011/7412765/Zone-revaluations-deplorable; The EQC "waiting for us to die," Charles Anderson, *Christchurch Press*, March 15, 2014; Why is Alf still waiting, 92 year-old pensioner stuck in damaged home, Ashleigh Stewart, *Christchurch Press*, June 7, 2014.

micromanage the free flow of information and decision-making processes. They are slow to act, wanting to take control of all decisions and often have external agendas. The New Zealand has no "upper house" of government to provide balance in the democracy, and it is this single-chamber political system that decides through its public policy process what recovery will look like in many of the decisions, actions, or inactions it chooses to pursue. Yet, any initial disaster and recovery response *must* be focused at the local level, because of local intimate knowledge of the environment in which the emergency occurs and of the immediate and future nature of community needs. Without local level "buy-in," recovery will be slow and the maintenance of public confidence among decision-making bodies is critical. Local knowledge has only sparsely been mobilized, and as a consequence recovery *has* been slow in Christchurch. In the wake of those policies in present capitalist society, which leave recovery to the market, a series of human rights violations have also taken place as a result of government failure to protect the interests of its citizens over and above the financial economic interests of corporates and big business.[29] If governments decide to leave a recovery "to the market," then it is the population that carries the weight of that policy of noninterference. In Christchurch these circumstances have come to the fore in the six examples discussed further.

THE EQC[30]

The earthquakes have depleted the natural disaster fund that underpins the EQC. In September 2011, the High Court of the New Zealand determined that EQC had financial exposure[31] after each major earthquake. Many believe that these financial pressures have much to account for the delayed claim payments and slow land damage assessments. In the past, EQC did not participate in the emergency response or recovery to disasters nor had it envisaged multiple events within such a short timeframe. EQC's operating environment changed markedly after the 2010–11 earthquakes, and it now has the responsibility for many recovery functions. EQC has come under

[29] http://www.stuff.co.nz/the-press/business/the-rebuild/67535901/New-Brighton-neglect-is-moral-injustice-Paul-Zaanen; http://www.stuff.co.nz/the-press/news/70087549/Red zone-discrimination-against-uninsured-says-city-council; https://www.hrc.co.nz/your-rights/social-equality/our-work/canterbury-earthquake-recovery/; Authorities "ignore" most Maori post-quake, Shelley Robinson, *Christchurch Press*, September 23, 2014; http://www.stuff.co.nz/national/health/76060300/Attempted-suicides-highest-in-Canterbury-twice-as-much-as-Auckland; The EQC bosses meet Brownlee over plight of elderly victims, Fairfax NZ, *Christchurch Press*, March 8, 2014; Children the first victims, Olivia Carville, *Christchurch Press*, February 16, 2013; Quake-hit families still in squalor, Olivia Carville, *Christchurch Press*, February 29, 2013

[30] http://www.eqc.govt.nz/.

[31] The New Zealand High Court was asked to resolve the issue of how EQC cover responds to homeowners who have made more than one claim for damage suffered in more than one earthquake, where such damage exceeds, NZ$100,000 for dwellings (or NZ$20,000 for contents). It was determined by the court that if, at the time of the subsequent earthquake, EQC had not yet paid in respect of the first earthquake, EQC is liable to pay up to NZ$100,000 with respect to each earthquake, until cover is reinstated. See https://www.interest.co.nz/sites/default/files/Earthquake%20Commission-1.pdf.

harsh criticism from Canterbury citizens regarding its governance structure, lack of transparency, and operating systems. Despite this, it continues to fail to accept any responsibility or accountability for the harm it has caused as a result of systematic assessment discrepancies, delays caused by reassessment (apportionment), failure to pay tradespeople in a timely fashion, poor workmanship requiring re-repairs, repairs carried out without building consents, the use of introduced building standards (the Ministry of Business, Innovation, and Employment Guidelines) as a way of cost saving, failure to address mold and asbestos risks adequately, etc. EQC on behalf of the government is widely seen to be involved in a government-driven cost saving exercise at the expense of affected earthquake victims.

SOUTHERN RESPONSE[32]

On April 7, 2011, it became clear that the insurance company AMI was in difficulty. AMI had more than 30% market share of the fire and general insurance market in Canterbury. The government assisted AMI with its financial difficulties, and Insurance Australia Group (IAG) agreed to purchase AMI, which the government declared was to "strengthen the Canterbury insurance market and reduce the Crown's liability."[33] The government-rolled AMI's liabilities into a government entity (Southern Response) set up with the sole purpose of settling the remaining AMI claims.[34] A group action is now in place against Southern Response, and proceedings allege that Southern Response misrepresented the terms of the original policy, and that Southern Response is only meeting 40%–50% of the amounts claimed or due to be claimed, based on its own inadequate building reports. Southern Response is seen to be concerned solely with limiting its own financial exposure. Being a government entity, it is seen to be trying to save money (tax payer's money) at the people's expense. When it appears that a government is experienced at cheating its citizens, the state of the democracy is rightly questioned.

THE PRIVATE INSURANCE INDUSTRY

The slow progress in recovery has been in large part due to the private insurance industry and its lack of willingness to process claims within reasonable timeframes. Earthquake insurance is known to be particularly problematic (Kunreuther & Roth, 1998, pp. 97–124). There are still thousands of unresolved insurance claims and many disputes, meanwhile insurance profits soar.[35] Now, into the sixth year after earthquakes, sound prudential management of the industry is and was required, bearing in mind the need to restore economic activity and the need of citizens to return to

[32] http://southernresponse.co.nz/.

[33] IAG agreed to purchase AMI, which the government declared was to "strengthen the Canterbury insurance market and reduce the Crown's liability."

[34] http://www.stuff.co.nz/the-press/business/71470352/southern-response-class-action-launched.html.

[35] http://www.stuff.co.nz/the-press/business/9000691/New-Zealand-disasters-boost-insurers-profits; http://www.interest.co.nz/business/68631/iag-posts-755-gain-profits-ahead-takeover-lumley.

their homes. In addition, the insurance industry in the New Zealand is "self-regulating." The emphasis is on "freedom, independence, and autonomy" with no or little government interference. This is sometimes a very convenient place for governments to sit. Yet, no freedom is unrestricted unless controlled by "accountability." For self-regulation to be effective, it needs to be properly integrated into the overall regulatory framework—that is, it needs to dovetail with the law and the regulator's policies and include regular direct oversight of the activities of the "profession." Central government intervention was and is required in light of what has transpired in Canterbury with regard to the insurance industry (Miles, 2016).

Questions also arise around the influence of reinsurers[36] on government. A nation that is unable to finance its own natural disasters finds itself at the beck and call of the reinsurer, being careful not to step out of line for fear of being without catastrophe protection. One can only wonder about the conversations that have taken place behind closed doors (such as the New Zealand delegation rendezvous in Monte Carlo in September 2011), which ultimately have an enormous impact on a nation's citizens.[37]

THE RED ZONE AND FORCED MIGRATION

The red zone[38] is one of the major issues that arose as a consequence of government/CERA "pay-outs" on properties at property values based on 2007 quotable value[39] figures in Christchurch and its environs. The Minister of Earthquake Recovery disregarded advice from officials to give red-zoned Cantabrians full compensation for their quake-damaged properties. Instead, only half the ratable value was offered. These property values were then already 5 years old and unrevised. For many citizens, this package was not sufficiently equitable to allow them to buy an equivalent property elsewhere. In April 2012, CERA ruled out a review of the residential red zone despite homeowners challenging the decision to "write-off" their land. A group of red zoners fought the government's "abuse of power" said to be "oppressive, disproportionate [and] contrary to human rights" in the courts.[40] This case marked a

---

[36] Reinsurance is most simply described as insurance for insurance companies. It occurs when multiple insurance companies share risk by purchasing insurance policies from other insurers to limit the total loss the original insurer would experience in case of disaster. By spreading risk, an individual insurance company can take on clients whose coverage would be too great of a burden for the single insurance company to handle alone. When reinsurance occurs, the premium paid by the insured is typically shared by all of the insurance companies involved.

[37] http://www.stuff.co.nz/the-press/opinion/perspective/5665147/Brownlee-gives-reinsurers-the-facts.

[38] Being "red zoned" means that the land has been so badly damaged by the earthquakes. It is unlikely that it can be rebuilt on for a prolonged period. The criteria for defining flat land areas as residential red zone are where there is significant and extensive area wide land damage; the success of engineering solutions may be uncertain in terms of design, success and possible commencement, given the ongoing seismic activity. Any repair would be disruptive and protracted for landowners.

[39] Quotable Value Limited is a state-owned enterprise that has the function of establishing land values for local authority rating purposes.

[40] http://www.stuff.co.nz/national/8946555/Complex-red-zoner-High-Court-case-opens.

test of primary property rights in the New Zealand. The claimants sought a judicial review of the government's compensation policy for red-zoned land and won their case right through to the Supreme Court. The court found that the government's red zoning policy had left landowners out of pocket and ruled that the government had not properly considered the Canterbury Earthquake Recovery Act and its purpose of "social, economic, cultural, and environmental well-being" when making the reduced offer.[41] More likely than not, one speculates that this land, now government land, will be made available once again for privatization.

THE ANCHOR PROJECTS[42]

NZD158 million was earmarked by the government for 16 multimillion dollar "anchor projects" that include a convention center, a rugby stadium, and metro sports facility. CERA wished to acquire central city land for these anchor projects. The plan included buying some 840 "designated" properties in the central city, condensing the city's core, and creating a more commercially attractive center, streamlining development rules and dividing the city into precincts. The government had a rigid vision wanting an IT and innovation hub, underpinned by a master plan complete with a network of laneways weaving through a "public realm" space. In addition, it wanted the right to approve prospective tenants. Landowners in the city stated that the Christchurch Central Development Unit was making offers to them that are "off the planet." The offers were said to be below some people's mortgages and well below market value. In total, 92 properties were to be acquired.[43] Meanwhile, hundreds of damaged community facilities are said to be likely to go unrepaired due to the CCC's shortfall in funding.[44] The Mayor, Leanne Dalziel, declares that tough calls will have to be made on which community facilities in the city are to be rebuilt.[45] While people are still living out of their homes, many feel that these projects are an extravagance established too early in the recovery phase. The public urges restraint on big projects. A trade-off in decisions has been developed between spending on "anchor projects" and making sure that the city's infrastructure is sound. The community questions whether Canterbury is seeing "disaster capitalism" in progress (Klein, 2009).

DEREGULATION IN THE FORM OF RELAXED BUILDING REQUIREMENTS (MBIE GUIDELINES)

Before the Canterbury Earthquakes, the New Zealand construction industry was regulated by the Building Act 2004 and the associated Building Code. Soon after the earthquakes, the Ministry of Business, Innovation and Employment (MBIE) produced a set of MBIE guidelines "*Revised guidance on repairing and rebuilding houses affected by*

[41] http://www.stuff.co.nz/the-press/news/christchurch-earthquake-2011/67321207/Quake-Outcasts-win-in-Supreme-Court.
[42] http://www.futurechristchurch.co.nz/central-city.
[43] Bitterness at CCDU valuations, Georgina Stylianou, *Christchurch Press*, July 13, 2013.
[44] Cash runs short for repairs, Lois Cairns, *Christchurch Press*, May 8, 2014.
[45] NZD 30m aquatic center on hold, Lois Cairns, *Christchurch Press*, February 27, 2014.

the Canterbury earthquake sequences (November 2011)," indicating what would be considered acceptable repair strategies post-earthquakes. These guidelines were introduced to encourage "speedy recovery."[46] The guidelines, adopted by EQC, have been used by builders, assessors, and engineers when qualifying and quantifying earthquake damage. As the years have progressed, the MBIE guidelines have had several revisions, each time to a lesser standard and further from the legislative national building standards. As the "goal-posts" for acceptable levels of damage have continued to move, the EQC's and private insurers' reassessments of property also moved, with their financial liability inevitably trending in a downward direction. Although the guidelines may be indicative of possible solutions to damaged dwellings, they cannot and do not override the provisions of any private insurance policy held with an insurer where damage is deemed to be above the quantum covered by the EQC,[47] nor can they override the requirements of the New Zealand Building Act 2004 and Codes. However, the private insurance industry has conveniently jumped on the bandwagon and also extensively used the MBIE guidelines and the expanded tolerances set out within these guidelines as justification for repairs which often, in their view, do not require building consents and which are treated as repairs rather than rebuilds. This ultimately saves the insurance industry hundreds of millions of dollars at the expense of the future integrity of building structures, and ultimately, the equity in and safety of policyholders' homes. "Cantabrians are being asked to wear second-best fixes because anything else is too expensive."[48] Today, the existence and intransigence of corruption in the construction and engineering industry is widely and publically acknowledged in Christchurch amid affected residents—ultimately should depend on local authorities—design permissions, foundations permissions, and occupancy permission. What has followed is a disproportionate number of newly built structures with building defects and repairs carried out shoddily, resulting in a poor quality housing stock as a result of a lack of building consent processes. Despite the government's awareness of these issues, it has done nothing to remedy the problems and knowingly allowed both EQC, Southern Response, and the private insurers to cut costs at the expense of the future building stock of the city and the equity in people's homes.

CONCLUSION

The primary focus by the national government with respect to the financial/economic long-term recovery of Canterbury puts the recovery of its population as a secondary concern. This has affected the overall well-being of many residents to a point that their human rights are seriously challenged. It represents an abdication of responsibility for the ensuing social disaster as well as permissiveness toward serious human rights violations.

[46] https://www.building.govt.nz/building-code-compliance/canterbury-rebuild/repairing-and-rebuilding-houses-affected-by-the-canterbury-earthquakes/; https://thechristchurchfiasco.wordpress.com/2015/03/01/mbie-guidelines-sarahs-view-on-the-matter/.

[47] EQC cover effectively operates as a "first layer," with the upper layer, i.e., damage that exceeds the first layer known as "the cap," covered by the private insurer.

[48] Who can we trust, John McCrone, *Christchurch Press*, March 23, 2014.

Early on in the process, Canterbury was judged by central government to be unable to manage a disaster of this magnitude alone. For this reason it was decided that the disaster needed a stronger command and control system than could be provided locally. CERA was put in place to oversee the recovery. The fact that CERA has become increasing unpopular is because it has demonstrated little understanding of community development, little empathy or comprehension of what is needed to build on the resilience of the people and the social capital available. When comments are made to the extent that not all is well in Canterbury, the cries of distress have been/are minimalized by referring to the earthquakes as "an unprecedented event" or "one of the largest insurance events in the world." In international literature one will find an emerging consensus that the first priority of a government after a natural disaster should be the facilitation of the reestablishment of communities (Brigit & Hagan, 2007). To prevent people from leaving the area, government should empower local initiatives to keep local communities "in the loop" by bolstering social cohesion. A quick return for residents to their homes should be a prime goal in the recovery phase and if that is not possible, provisions such as temporary and transitional accommodation should be in place.

As demonstrated, declaring a disaster a "national emergency" has profound political implications. In the follow-on from managing the emergency and the rescue efforts, it is practically unavoidable that a further politicization of the event increases as the affected community moves from the emergency response through the recovery and the reconstruction phases. The immediate emergency response by any government is fairly predictable, as it should be, but, from a political point of view, the aftermath has proven to be uncharted territory, highly susceptible to the opportunities of the circumstances, and the political values and agendas of the day. The way a government perceives its political mandate, or is given opportunity to define it, is never more critical than in a recovery phase. The malfeasances of the private insurance industry and the construction industry as described above are cases in point.

Markets have no inherent moral character, and it is therefore arguable that it is the government's role to decide how to manage them. In particular, after a major disaster, markets must be regulated "under emergency" to ensure that they are working for the benefit of the recovery of the majority of citizens. A political system of noninterference only serves to amplify the voice of wealthy corporates and fails to protect the ordinary citizen against corporate abuse. Money speaks in politics as it does in the marketplace. Any system of recovery must have rules and regulations operating within a legal framework. In a modern economy, the government has the responsibility on behalf of its population to set and enforce the rules of the game in the marketplace. This is especially true in the case of a major disaster where government takes the decision to be involved in the recovery process. In the absence of genuine government support, the extent to which a population can recover post-disaster is likely to be severely challenged.[49] What has characterized the recovery

[49] http://www.stuff.co.nz/the-press/news/christchurch-earthquake-2011/9397762/Cunliffes-recovery-criticism-ridiculous.

in Christchurch is that political decision-making has been in favor of the corporate and government stakeholders—the insurance industry and the construction industry. The policy of noninterference in the marketplace has been the cause of slow, painful recovery. The consequences of this approach have been sorely and visibly felt by the affected population. In many instances this has resulted in a series of serious misunderstandings about the way forward between national government, the CCC, and its population. It is imperative that government has a directive role on how the industry intends to operate postdisaster, and therefore it must communicate more effectively with the parties involved and align policies and implement strategies together. Without a comprehensive alignment in policy and the allocation of responsibilities and the development of a regulatory framework, those directly affected by a disaster will continue to be victimized long after the event itself. Had government goodwill been forthcoming and this preparatory work taken place, or been an already established part of policy/legislation and understood by all parties, the timeframes for citizen recovery would have been shortened, particularly in relation to insurance claim resolution even in the absence of legislated standards for the industry.

Confidence and certainty around citizens' futures in a devastated community is vital in the recovery phase. After 6 years the continued delays in insurance settlements have led to widespread uncertainty, causing unrest and desperation among a few.[50] This development in itself will ultimately have an enormous impact on how those individuals perceive their city and their government. The public perception of clear and authoritative leadership whether in planning or in housing assistance is important and requires special attention at any time. In this respect, there still remains a mismatch between residents' expectations and the decisions that are made by CERA and other governing bodies including the CCC. There is a need to build transparent governance, citizen/state trust. Where citizens feel the most distance from decision makers, these are the communities that experience the slowest and the most confused of recoveries (Aldrich, 2012). Societies where citizens can access decision makers and processes have their opinions heard, and where required, alter state policies to a significant degree, are societies in which recovery will take place more smoothly. There is a close link between the ability of citizens to have their voices heard and broader trust in their government. In a society where citizens have low efficacy and believe that their collectively accepted ideas and desires are not permeating into actual policy, there will be a diminishing trust in government. This has its immediate effect on the duration of recovery, and, unfortunately the recovery of Christchurch after the 2010–11 earthquakes demonstrates this clearly. In that respect the story of Christchurch is a story of a missed "window of opportunity" as governmental financial interests took precedence over public well-being in the interests of profits and cost-savings.

[50] http://www.stuff.co.nz/the-press/business/the-rebuild/78233527/Housing-repairs-will-create-hideous-conditions.

REFERENCES

Aldrich, D. P. (2012). *Building resilience: Social capital in post-disaster recovery*. Chicago: University of Chicago Press.

Alexander, D. E. (2002). The evolution of civil protection in modern Italy. In J. Dickie, J. Foot, & F. Snowden (Eds.), *Disastro! Disasters in Italy since 1860, in: Culture, politics, society* (pp. 165–185). New York: Palgrave Press.

Alexander, D. E. (2008). *Principles of emergency planning and management*. New York: Oxford University Press. Harpenden, UK.

Berke, P. R., Kartez, J., & Wenger, D. (1993). Recovery after disaster: Achieving sustainable development, mitigation and equity. In *Disasters* (Vol. 17) No. 2.

Blaikie, P., Cannon, T., Davis, I., & Wisner, B. (1994). *At risk: Natural hazards, people's vulnerability, and disasters*. New York: Routledge.

Brigit, M., & Hagan, P. (2007). Disasters and communities: Understanding social resilience. *Australian Journal of Emergency Management*, *22*, 16–20 May 2007.

Brinkerhoff, D. W. (1999). *Identifying and assessing political will for anti-corruption efforts*. Working Paper No. 13, Implementing Policy Change Project. Washington, D.C.: U.S. Agency for International Development, Center for Democracy and Governance.

Coppola, D. P. (1999). *Introduction to international disaster management*. Oxford: Butterwoth-Heinemann.

Coppola, D. P. (2006). *Introduction to international disaster management* (3rd ed.). Oxford: Butterwoth-Heinemann.

Klein, N. (2009). *The shock doctrine: The rise of disaster capitalism* (6th ed.). New York: Metropolitan Books/Henry Holt.

Kunreuther, H., & Roth, R. J. (1998). *Paying the price: The status and role of insurance against natural disasters in the United States*. Washington, DC: Joseph Henry Press.

Miles, S. A. (2016). *The insurance aftershock: The Christchurch Fiasco post-earthquakes 2010–2016* (2nd ed.). New Zealand: Labyrinth Publishing.

Millen, D. (2011). *Deliberative democracy in disaster recovery: Reframing community engagement for sustainable outcomes*. Sydney: University of Western Sydney.

Nakagawa, Y., & Shaw, R. (2004). Social capital: A missing link to disaster recovery. *International Journal of Mass Emergencies and Disasters*, *22*(1), 5–34.

Pelling, M. (2003). *The vulnerability of cities: Natural disaster and social resilience*. U.K.: Earthscan Publications Ltd.

Rotimi, J. (2010). *An examination of improvements required to legislative provisions for post disaster reconstruction in New Zealand*. Ph.D. thesis. Christchurch: University of Canterbury.

Schneider, S. K. (March–April 1992). Governmental response to disasters: The conflict between bureaucratic procedures and emergent norms. *Public Administration Review*, *52*(2), 135–145.

Schneider, S. K. (1995). *Flirting with disaster: Public management in crisis situations*. New York.

Vallance, S. (2011). Early disaster recovery: A guide for communities. *Australian Journal of Disaster Recovery & Trauma Studies*, *12*, 19–25.

FURTHER READING

[1] Brookie, R. (2014). Governing the canterbury earthquake recovery, 2010–2011: The debate over institutional design, Part 4, Chapter 20. In J. Boston, J. Wanna, V. Lipski, & J. Pritchard (Eds.), *Future – Proofing the state: Managing risks, responding to crises and building resilience* (pp. 251–276). Australian National University Press.

[2] Connolly, M. J. (2013). *The impacts of the Canterbury earthquakes on educational inequalities and achievement in Christchurch Secondary Schools.* Master of Science Thesis (Geography). University of Canterbury.

[3] Constable, S. T. (2013). *Red zone as green corridor: Opportunities for suburban intensification: A christchurch case study.* Master Thesis. Victoria University of Wellington.

[4] Gordon, L. (April 2015). Rich and poor schools revisited. *New Zealand Journal of Educational Studies*, *50*(1), 7–21.

[5] Mannakkara, S., & Wilkinson, S. (2014). Re-conceptualising "Building Back Better" to improve post-disaster recovery. *International Journal of Managing Projects in Business*, *7*(3).

[6] Miley, F., & Read, A. (2013). After the quake: The complex dance of local government, national government and accounting. *Accounting History.*

[7] Mitchell, D., Myers, M., & Grant, D. (2015). Land valuation: A key tool for disaster risk management. *Land Tenure Journal.*

[8] Taylor, J. E. (2013). *The Christchurch Earthquake Sequence: Government decision making and confidence in the face of uncertainty.* Thesis Master in Arts. Queen's University.

[9] Toomey, E. (2011). The slow road to recovery: A city rebuilds under the Canterbury Earthquake Recovery Act. In S. Butte, H. Nasu, & L. Nottage (Eds.), *Asia-pacific disaster management: Comparative and socio-legal perspectives.*

[10] Vallance, S. (2013). *Engaging communities, Waimakariri district council's integrated community-based recovery framework.* New Zealand: Faculty of Environment, Society and Design Lincoln University.

CHAPTER

From Recovery to Prevention: The Swiss Avalanche Program

7

Maria Kornakova[1,2], Alan March[1]

[1]The University of Melbourne, Melbourne, VIC, Australia; [2]Massey University, Palmerston North, New Zealand

INTRODUCTION

Rapid urbanization and climate change are strongly aligned with the likely increase in numbers of natural disasters in the near future (Wamsler, 2014). Therefore, there is a strong need to realistically assess them and find ways to improve existing settlements or develop new prevention mechanisms to minimize the negative impacts of disasters. Typically, all activities, programs, and measures occurring before, during and after a disaster event are addressed through disaster risk reduction (DRR) practices, which aim to avoid or minimize disaster impacts, and aid the recovery of affected areas (Vasilescu, Asmatullah, & Himayatullah, 2008). These activities, although often intertwined, generally follow the cyclical nature of a disaster and addressed in the four phases or stages: Prevention, Preparedness, Response, and Recovery (PPRR) (e.g. Clary, 1985; Godschalk & Brower, 1985; Haddow, Bullock, & Coppola, 2011; Mileti, 1999; Quarantelli & Kreps, 1972). Disaster events occur between the preparedness and response stages, both of which are short term insofar as they are oriented to response actions in the short term—even while preparation itself might occur on an ongoing basis. The recovery stage, however, includes short- and long-term actions, including those related to urban planning. Immediate recovery also includes the so-called "window of opportunity" stage, allowing professionals and specialists from the field to intervene in existing practices and perhaps to implement more appropriate and far reaching measures that may not have otherwise been palatable. The desire of the affected stakeholders to return to the "normal" state of operations as soon as possible usually limits the timeframe of the window of opportunity, and, as a result, recovery actions often lead to rebuilding of affected areas in a manner similar or the same as the predisaster stage (Haigh & Amaratunga, 2011). However, active and effective use of the window of opportunity has the potential to address the need to improve or change existing prevention mechanisms.

This chapter provides an insight into effective employment of the "window of opportunity" using the case of Swiss avalanche practice. For the context of this chapter, only snow avalanche is discussed, which is understood as a large snow or

Urban Planning for Disaster Recovery. http://dx.doi.org/10.1016/B978-0-12-804276-2.00007-4

rock mass rapidly moving down a mountainside. It is usually triggered by heavy snowfall, rain, defrost, or physical disturbances, and contains rocks, soil, or ice (Alexander, 1999; Chapman, 1999; McClung & Schaerer, 1993; Schweizer, Bruce Jamieson, & Schneebeli, 2003). Snow avalanches are common in the populated parts of mountain regions or in the regions with increased winter levels of activity, such as tourism. Compared to some other natural disasters, the area of the potential avalanche formation can be predicted relatively accurately as they require critical snow mass and sufficient slope for the movement. As such, the most dangerous slopes are in the range of 25°–40 as they allow collection and movement of the critical snow mass (Alexander, 1999; International Association of Hydrological Sciences & International Commission on Snow and Ice, 1981; McClung & Schaerer, 1993).

Switzerland is the leading country in avalanche prevention and response programs. While the unique topography of this country results in the presence of a number of hazards, the combination of unique topography, high density, and high demand for the mountain tourism places avalanche as the major hazard in the country (Federal Statistics Office, 2012; Gillet et al., 2006; Laternser & Schneebeli, 2002). The history of the first recorded avalanche in Switzerland dates back to 1449, when—according to the available data—this natural hazard destroyed four buildings and killed 11 people. Over the centuries, such records became more frequent and detailed (Schneebeli, Laternser, Föhn, & Amman, 1998), allowing for more accurate studies, and subsequently, more accurate predictions of this hazard (Frutiger, 1970, 1980). Recently, snow avalanches, landslides, debris, and rock falls have become more frequent and with higher amplitude, possibly due to the increasing development activities, forestry practices, land misuse, and climate change in the mountain areas (Pudasaini & Hutter, 2007), indicating a need for greater attention to, and additional measures against, avalanches.

This analysis follows particular events of the avalanche season of 1951, also referred to as "Winter of Terror." These events were chosen as a trigger point for changes in the avalanche prevention approach in the country, which revealed the complexity of agents involved in and forces influencing decision-making, leading incrementally to the changes to hazard planning over the following decades. The chronological overview presented of the events leading to and following the "winter of terror" describes the governance associated with the integration of urban planning and DRR disciplines, and indicates the importance of the roles of stakeholders in the processes. This unique approach and general cultural acceptance of the role of authority in Switzerland is suggested as one of the reasons for stakeholders' behavior, as well as demonstrating the potential of clearly established governance in influencing community behavior. This case also demonstrates how economic imperatives influence residents' appetite for change and their propensity to modify their attitude toward enforced avalanche zoning. Data for this are collected through secondary sources by using document analysis. In addition, upon the initial write-up of the case, semistructured interviews with key professionals in the field and observations were used to fill in gaps in the data and confirm or enhance the initial findings of authors.

LAND USE CHANGES IN SWITZERLAND IN THE FIRST HALF OF THE 20TH CENTURY

A shift from an industrial to a mainly service-based economy in Switzerland in the early 20th century and post World War II, combined with its climatic conditions and topography, resulted in rapidly developing tourism, particularly in winter sports such as skiing (McClung & Schaerer, 2006). This shift resulted in decreasing land values for rural purposes in connection with decreasing farming activities in the country in the first half of the 20th century and subsequent mass movements of farmers to larger cities and urbanization, resulting in loss of land values in mountain areas. At the same time, the rapid growth of mountain tourism attracted foreigners and investors to the country. Vacant land has been sold to private owners with the main purpose of building second or holiday houses. The majority of new buyers were foreigners, who had inadequate knowledge of avalanche threats in the area, which attracted land speculation. Moreover, as disastrous avalanches have relatively long return intervals (McClung & Schaerer, 1993), the "living memory" of such events faded and even local residents were not always aware of possible threats. As a result, some of the newly purchased land and constructed houses were in avalanche risk areas (Frutiger, 1970, 1980).

The situation came to a head in 1951, when a series of avalanches occurred across the Alps and resulted in a significant damage to the whole region. Also known as the "Winter of Terror", the season of 1951 brought two catastrophic avalanche cycles to the Alps region. Excessive snow levels lead to 649 avalanche events across the whole region, causing severe economic and environmental damage with deaths of 98 people in Switzerland alone and 265 people across the entire region (Frutiger, 1970, 1980; Gilg, 1985; Lateltin & Bonnard, 1999; Margottini & Casale, 2003; Swiss Disaster management Professional, 2013; The Swiss Federal Institute for Snow and Avalanche Research, n.d.; Wilhelm, Wiesinger, Bründl, & Ammann, 2001).

Impact of these events led to conflicts between landowners and public authorities responsible for the welfare of inhabitants, which includes damage from avalanche events (Frutiger, 1970). Conflicts resulted in dramatic increases of both civil lawsuits between sellers and buyers, and public legal actions between landowners and local authorities (Frutiger, 1970, 1980; Lateltin & Bonnard, 1999). The reaction of the public and local authorities was expressed through increased appetite for change; yet, there were no immediate actions from the government at that time. Instead, federal government issued additional guidelines that were intended to provide for the application of the Federal Acts of December 6 and 19, 1951 (Frutiger, 1980). Aimed at increased effectiveness of avalanche defense mechanisms, these acts did not include avalanche zoning and remained silent regarding the land speculation rife in regions prone to avalanches. They discussed the need for additional help in regions affected by the Winter of Terror with afforestation and avalanche defenses (Frutiger, 1980).

Upon development and implementation of the first avalanche plan by the SLF for the community of Wengen, Canton of Bern (The Swiss Federal Institute for Snow and Avalanche Research, n.d.), the first avalanche zoning map was developed in 1961. As a response to the growing avalanche risks, the Federal Bureau of Forestry raised the need for avalanche zoning plans for the communities prone to this disaster. They proposed insurance as a tool restricting the growing numbers of developments in areas with high avalanche risks. While the political structure of the country did not allow confederation to enforce these rules upon cantons due to the political structure of the country, it still released relevant guidelines in article 32 of Executive Ordinance to the Federal Forest Law of October 1, 1965 (Frutiger, 1980). This tactic also illustrates how economic incentives can be used to change the development patterns in risk adverse areas. It also highlights the fundamental importance of support being provided by the background rules of the system and decision makers in achieving integration processes of DRR and urban planning.

LAT[1] AND LAND USE CHANGES

Land speculation continued to occur at a greater rate over the following 18 years, forcing cantons to request intervention from the confederation. The main challenge that impeded efficient problem solving was significant differences of building laws and codes between different cantons (Frutiger, 1980, p. 319). This is because Switzerland is a federation with 26 cantons or states. Central authorities have jurisdiction only in the domains outlined by the federal constitution while remaining powers automatically revert to cantons or communities. Cantons are sovereign, and each has its own government, which is a constitutional entity issuing laws and regulations in accordance with the framework outlined and defined by federal laws. Management and prevention of all natural disasters, including avalanches, follows the same regulations (Lateltin & Bonnard, 1999).

The avalanche disaster events affecting the country in 1968 and 1970 resulted in a number of interventions and disputes being taken to the parliament, resulting in postulates and discussions between officials about the need of reducing risks in the development areas. At that time, the 12 cantons and communities across the country had already started making progress toward introducing avalanche zoning (Frutiger, 1980), demonstrating the power and benefits of grassroots approaches.

As a free enterprise society (Gilg, 1985, p. 319), the Swiss Constitution establishes that each and every property owner, both private and public or communal, has absolute power over their land and its development. Until 1969, this principle was stated in the first paragraph of the constitution and was obstructing the progress of planning practice in the country. In 1969, when the 22nd article was restated, it provided power "to the confederation or canton to make a provision for the expropriation or restriction of land ownership with compensation" (Gilg, 1985, p. 319). This was a first step toward the

[1] Loi sur l'aménagement du territoire.

introduction of zoning and planning principles. The level of local autonomy provided by these articles allowed for place and context-specific decisions to be made by local people and agencies, replacing wider scale maps that were more generic and not necessarily applicable to the area, which were developed by national officials (Frutiger, 1980).

The first draft of the Federal Act, highlighting the need for urgent changes in land use planning, was presented by the Federal Council in 1972 (Gilg, 1985). This draft responded to the need or landscape protection, which did not include the avalanche hazard zones at this time. Upon being heard in the Federal Assembly, this deficiency was addressed and the law stated that:

> *Some organizations considered it advisable to pay attention, in addition to environmental protection, to the restriction of building activity in areas endangered by natural hazards. Some memorials would like to oblige the Federal Council to take precedence over negligent cantons and, if necessary, to designate the areas to be protected and act in their place.*
>
> **Switzerland (1972, pp. 644–648)**

Furthermore, the law ordered all cantons to urgently designate "e) areas which are known to be endangered by natural hazards" and emphasized that "… the Federal Council will, after unheeded warnings, itself take measures against cantons, communes or other disloyal bodies" (Switzerland, 1972, pp. 644–648). These measures were enforced for a limited time and were replaced by the permanent Federal Law for Land Use (LAT) (Frutiger, 1980).

The Federal Forest Law and the Federal Law on Flood Protection introduced and applied across all cantons in 1991, places emphasis on preventive measures for natural hazards, including avalanche (Lateltin & Bonnard, 1999), establishing land use planning as central tool alongside hazard assessment itself. Following these laws, cantons were required to develop maps and registers for endangered areas and consider hazards when land use guidelines are being established (The Federal Assembly of the Swiss Confederation, 1991). These are supported by LAT that requires all cantons to develop a master plan that must include hazard mapping relevant for each individual canton. It is further scaled down to the communal level, and detailed plans specific to the place and hazard type are further requested by local authorities from communes (Frutiger, 1980; Lateltin & Bonnard, 1999). In areas exposed to avalanche hazard, such maps are resented in the avalanche zoning tool.

AVALANCHE ZONING—PURPOSE AND TECHNICALITIES

Avalanche zoning, or avalanche zone plans, is a prevention tool, enforced by the Swiss law and regulated by local officials. This tool identifies avalanche risk levels and applies relevant land use planning restrictions when required. Following principles of land use planning, it places restrictions based on spatially defined zones and specifies types of construction allowed in these zones (Frutiger, 1970; McClung & Schaerer, 1993; WSL Institute for Snow and Avalanche Research SLF, n.d.-b).

Areas exposed to the hazard are based on avalanche mapping—large topographic maps outlining potential areas and paths of this snow hazard. This mapping is a tool of building authorities, and it does not have a direct legal status (Frutiger, 1980), whereas zoning does have a legal status.

Avalanche zoning is an example of the integrated efforts of diverse range of professionals, which was developed and implemented as an outcome of the long-term recovery process initiated after the disastrous 1951 events. Its development consists of a number of steps, understanding of which is important for the understanding of the overall process and for the discussion provided further on the role of citizens in the process. The first stage is the hazard assessment, in which the potential avalanche path is identified to recognize if and where any development, infrastructure, or other facilities are planned in mountainous terrain. Furthermore, these areas are subdivided into zones according to the risk levels, and finally, development restrictions are applied in these areas (McClung & Schaerer, 1993).

Avalanche zoning has four zones, each of which is assigned a color identifying danger level and development restrictions. The red zone is assigned to areas with the highest risks of avalanche and it bans all new development, requires reinforcement of all existing constructions and structures, and requires that evacuation plans and paths are ready at all times during the avalanche season. The blue zone allows some new development, but it must strictly follow regulated and standardized protection measures. Development type is also restricted in the blue zone to buildings not attracting large crowds of people, meaning that public facilities, such as schools, or mountain tourism amenities, such as lodges and lift terminals, are not permitted. The yellow zone has lower hazard levels and generally allows for all types of development with appropriate structural measures. The white zone has no avalanche danger and does not apply restrictions to the development (Frutiger, 1970; McClung & Schaerer, 1993; WSL Institute for Snow and Avalanche Research SLF, n.d.-a).

Despite a strong emphasis and enforcement of measures aiming to reduce risks associated with avalanches, by May 2013, at least one-fourth of hazardous areas in Switzerland were not assessed and treated by zoning in relevant maps. While Gilg (1985, p. 329) suggested that the initial 8 years provided for cantons to develop land use planning under LAT were rather challenging due to the lack of professionals, land speculations remain another potential reason for certain level of reluctance between local officials. The introduction of LAT and enforcement of hazard maps placed restrictions on land use and reduced land value. For example, as specified by the interview subject, the price of land per square meter in the least restrictive white zone in Davos in 2013 was about USD $2,000, while price per equivalent land in the most restrictive red zone was about USD $20–25 (Swiss Disaster Management Professional, 2013). Combined with the lack of adequate federal subsidies and cantons, this shift led to hesitation on the part of local authorities in introducing hazard maps (Frutiger, 1980; Lateltin & Bonnard, 1999). This example again demonstrates the importance of consideration of economic forces and a need for thorough understanding of all stakeholders and their vested interests on DRR processes, including urban planning.

ROLE OF STAKEHOLDERS IN MODERN AVALANCHE ZONING PRACTICES

The history of Swiss avalanche zoning and prevention tools and mechanisms demonstrates that successfully integrated DRR relies not only on the quality of prevention mechanisms but also on stakeholder involvement and managing productively their vested interests. Following Arnstein's ladder of participation, the discussion of community involvement demonstrates how after initially being the group to trigger change, the community accepted roles of being informed only.

By "informing," we refer to the ladder of participation first presented by Arnstein in 1969 and further amended and discussed by a number of scholars (Arnstein, 1969; Kloman & Arnstein, 1975). Table 7.1 is developed from a synthesis of Arnstein (1969) and the International Association for Public Participation (2003) as a referral point for concepts of participation used in this chapter It is suggested as a practical approach, as the spectrum of IAP2 (2003) appears to be most appropriate for the investigation being carried out in this research thesis. Arnstein's (1969) ladder adds additional critical descriptive levels of participation, which categorize instances in which government enforces opinion on citizens. Although such an approach does not allow participation for the citizens, it is assumed that in some planning cases such an approach might be appropriate, despite the difficulties associated with this. In addition, Table 7.1 includes some examples of common participation techniques.

As the Swiss Confederation requires an appropriate hazard map to be developed before statutory processes taking place, it is left to cantons to decide which hazard should be assessed and then mapped. Hazard maps are developed by professionals from relevant fields and further integrated in relevant practices, including urban planning. Only relevant professionals perform both processes and, as a result, updated hazard zoning is developed. On completion of zoning maps, local government makes decisions regarding their implementation. The relative freedom of officials and professionals to execute plans and zoning maps can be considered as empowerment being allocated to professionals and technical bureaucratic officials. As part of the implementation process, local governments and professionals work with potentially affected communities (Swiss Disaster Management Professional, 2013); however, the lack of a universal protocol for such work does not determine whether interests of landowners are considered in full, or whether instead an informative approach is taken and the consequences of not applying rules and regulations are provided to affected parties.

Despite there being greater communal power and authority to make significant changes in Switzerland compared to many nations, citizens are restricted in their rights to develop in the avalanche-zoned areas. Citizens can vote on the plan; however, their right to appeal against the hazard map itself is restricted to only those directly affected. The right of environmental agencies to intervene and appeal remains. Citizens' appeals are initially carried out at the canton level and, if case is not resolved, they are addressed at the confederation level. In either case, courts consult or refer back to urban planning and disaster professionals to seek for the

Table 7.1 Degrees of Participation, Their Goals, and Tools

Participation Degree	Goal	Method
Manipulation and therapy	Provide public with information on the chosen plan, no channel for feedback, aims to gain public support	• Public and community meetings • Public hearings
Informing	Provide public with adequate information to facilitate their education about existing problems and issues, alternatives, and solutions; public is informed	• Websites • Fact sheets • Local newspapers, newsletters • Progress reports, direct mail • Meetings • Public hearings • Surveys and questionnaires • Focus groups
Consult	Channel for feedback from public on analysis, decisions, and alternatives; opinion of public is considered but not necessarily included in the decision-making process	• Face-to-face interactions • Delhi process • Focus groups • Public meetings and hearings • Surveys and questionnaires
Involve	Work with public through the planning process to ensure clear understanding and consideration of their concerns and ideas, including opinions on the planning process, and informing public regarding the decisions	• Workshops • Brainstorming • Charretes • Games • Deliberative pooling
Collaboration or partnership	Creation of partnership with public in the decision-making process, direct advice from the public is incorporated in the decision-making process to the greatest possible extent	• Advisory committees • Consensus building • Participatory—decision-making • Policy communities
Empower	Final decision is given to public	• Ballots • Citizens' juries • Delegated decisions • Studies of impact assessment

Developed from Adams (2004), Arnstein (1969), Beutel and Dalton (2001), Bishop and Davis (2002), Healey (1997), Innes, 1995, 1996, 1998, International Association for Public Participation 2 (2000), Kloman and Arnstein (1975), and Sanoff (2000).

professional advice (Swiss Disaster Management Professional, 2013). As rights to appeal do not mean that any opposition will be effective on citizens' part but rather it is an opportunity for professional to react or reflect on the accuracy of mapping process, this can fluctuate between involvement and consultation participation types. This is of a particular interest as the initial trigger for change came from the community members.

However, the community is substantially involved in DRR processes on a daily basis. Citizens are informed about the risks of potential hazards in the area, and each household receives bulletins and hazard information; evacuation paths in the community are clearly marked and identified. Moreover, some local residents are involved in various emergency services such as fire brigades or rescue teams, and the community itself is the first response team in the case of a disaster event. If required, further aid is sought from a canton, and the confederation assistance is sought only when canton's forces are not efficient enough. Remaining highly successful (Swiss Disaster Management Professional, 2013), this practice demonstrates a more bottom-up approach, which occurs when a community is properly educated with regards to the broad spectrum of matters relevant to DRR.

Moreover, from citizens' perspective, an "informing" approach has been favored in the Swiss case over time, since the bulk of participation here is oriented to informing the entire community of proposed plans. There is no real redistribution of power, and option to appeal is limited only to those directly affected by hazardous zoning.

While some consultation processes can be observed at certain stages of development of maps, final decisions are yet to be undertaken by professionals. Despite what some might consider "false participation", this approach appears to be successful in this case and has reduced risks associated with avalanches, demonstrating a high level of acceptance of the role of professionals and their expertise by communities. While, in future, more in-depth research is required to identify all potential reasons for the roles of stakeholders, based on the data available a conclusion is made that a rather authoritarian approach to power distribution and clear regulation of responsibilities influences residents' acceptance of DRR measures. Moreover, it appears that a prolonged history of practice in the country has led to greater levels of acceptance of DRR tools being executed by the community members.

CONCLUSIONS

The history of the Swiss hazard mapping shows the potential for several dramatic events to influence increased awareness of potential disasters and to allow implementation of changes to relevant measures. As such, while the disaster events of the avalanche season of 1951–1952 in the Alps lead to significant changes in DRR practice of the country, two other disastrous events of 1968 and 1970 were necessary before parliament saw fit to intervene and restrict development in hazardous areas. Moreover, despite the Federal Law for Land Use (LAT) that was issued in 1978 requiring all Cantons to develop a Master Plan reflecting relevant hazardous territories, they still were not implemented completely by 2013–2014. This potentially demonstrates some level of resistance among officials and, perhaps, residents. It is proposed that there might be a range of potential reasons for this—political influence, areas of exposure, etc.; however, additional research is required to analyze and understand these in depth.

The Swiss case is an example whereby devolving control to the cantonal level and using experts' knowledge led to considerable certainty being reached regarding DRR outcomes, particularly with appeal processes allowing the possibility of mistakes to be identified and rectified. In other words, professionals with sufficient data and depth of understanding of threats and risks associated with disaster events were allowed to take actions significantly reducing risks. This case also demonstrates that an educative approach to residents and facilitation of their knowledge result in communities being rather resilient to disaster threats, suggesting favoring of this approach that privileges professionals over simple distribution of knowledge among residents. Moreover, response actions taken by the community and local officials demonstrate that community members are capable of coping with disaster situations and responding to them accordingly.

Avalanche zoning in Swiss areas appears to be highly formalized in its basis and has developed over the period of almost 30 years and tested favorably in several legal cases. It remains an example of regulatory planning as one of the main phases of implementation of necessary means for DRR. While cases leading to the legalization of avalanche zoning included a range of relevant professionals, legalization itself remained an urban planning, whereby hazard maps were added as a legal "bundle" in support to draw together key information and evidence.

While financial schemes are outside of the scope of this chapter, yet it is acknowledged that the Swiss case has the potential to be cost-effective, compared to other practices. It might to be less costly for government to include DRR through existing means and process and go through the approval process once, rather than issuing two separate documents and going through the approval processes twice. This also means that fewer expenses are put toward other related expenses, such as hiring facilities and staff. These, however, are assumptions requiring separate investigation.

REFERENCES

Adams, B. (2004). Public meetings and the democratic process. *Public Administration Review*, *64*(1), 43–54. http://www.blackwellpublishing.com/journal.asp?ref=0033-3352.

Alexander, D. (1999). *Natural disasters*. London: UCL Press.

Arnstein, S. R. (1969). A ladder of citizen participation. *Journal of the American Institute of Planners*, *35*(4), 216–224. http://dx.doi.org/10.1080/01944366908977225.

Beutel, L., & Dalton, A. (2001). Post-occupancy evaluation. *Australian Planner*, *38*(3–4), 178–181. http://dx.doi.org/10.1080/07293682.2001.9657965.

Bishop, P., & Davis, G. (2002). Mapping public participation in policy choices. *Australian Journal of Public Administration*, *61*(1), 14.

Chapman, D. (1999). *Natural hazards* (2nd ed.). South Melbourne, New York: Oxford University Press.

Clary, B. B. (1985). The evolution and structure of natural hazard policies. *Public Administration Review*, *45*(Special), 20–28.

Federal Statistics Office. (2012). *Population density, 2010*. Adobe Flash. Neuchâtel: Federal Statistics Office.

Frutiger, H. (1970). In R. Doebbling, & E. R. LaChapelle (Eds.), *The avalanche zoning plan*. U.S. Department of Agriculture, Forest Service, Intermountain Region, Alta Avalanche Study Center.

Frutiger, H. (1980). History and actual state of legalization of avalanche zoning in Switzerland. *Journal of Glaciology*, *26*, 313–324.

Gilg, A. (1985). Land use planning in Switzerland. *The Town Planning Review*, *56*(3), 315–338. http://dx.doi.org/10.2307/40112205.

Gillet, F., Pasqualotto, M., Raviol, P., Rouiller, J.-D., Requillart, J.-P., & Vengeon, J.-M. (2006). *The consideration of natural risk in land use planning*.

Godschalk, D. R., & Brower, D. J. (1985). Mitigation strategies and integrated emergency management. *Public Administration Review*, *45*(Special), 64–71.

Haddow, G. D., Bullock, J. A., & Coppola, D. P. (2011). *Introduction to emergency management* (4th ed.). Elsevier.

Haigh, R., & Amaratunga, D. (2011). A window of opportunity. *International Journal of Disaster Resilience in the Built Environment*, *2*(2). http://dx.doi.org/10.1108/ijdrbe.2011.43502baa.001.

Healey, P. (1997). *Collaborative planning: Shaping places in fragmented societies*. Houndmills: Macmillan Press.

Innes, J. E. (1995). Planning theory's emerging paradigm: communicative action and interactive practice. *Journal of Planning Education and Research*, *14*(3), 183–189. http://dx.doi.org/10.1177/0739456x9501400307.

Innes, J. E. (1996). Planning through consensus building. *Journal of the American Planning Association*, *62*(4), 460.

Innes, J. E. (1998). Information in communicative planning. *Journal of the American Planning Association*, *64*(1), 52–63.

International Association for Public Participation 2. (2000). *Public participation spectrum*. International Association for Public Participation. Retrieved from http://www.iap2.org/practitionertools/spectrum.html.

International Association for Public Participation. (2003). *The IAP2 public participation toolbox*. International Association for Public Participation. Retrieved from http://www.iap2.org/boardlink/toolbox.pdf.

International Association of Hydrological Sciences & International Commission on Snow and Ice. (1981). Avalanche atlas; illustrated international avalanche classification. *Natural Hazards*, *2*.

Kloman, E. H., & Arnstein, S. R. (1975). A working model for public participation. *Public Administration Review*, *35*(1), 70–73.

Lateltin, O., & Bonnard, C. (1999). *Hazard assessment and land-use planning in Switzerland for snow avalanches, floods and landslides* (Chapter 6) of the comprehensive risk assessment for natural hazards – a contribution to the international decade for natural disasters reduction (IDNDR). Available at: http://www.eird.org/deslizamientos/pdf/eng/doc12082/doc12082-contenido.pdf.

Laternser, M., & Schneebeli, M. (2002). Temporal trend and spatial distribution of avalanche activity during the last 50 years in Switzerland. *Natural Hazards*, *27*(3), 201–230. http://dx.doi.org/10.1023/a:1020327312719.

Casale, R., & Margottini, C. (Eds.). (2003). *Natural disasters and sustainable development*. New York: Springer-Verlag.

McClung, D., & Schaerer, P. A. (1993). *The avalanche handbook*. Seattle: Mountaineers.

McClung, D., & Schaerer, P. A. (2006). *The avalanche handbook* (3rd ed.). Seattle: Mountaineers.

Mileti, D. S. (1999). *Disasters by design: A reassessment of natural hazards in the United States*. Dennis S. Mileti; with the contributions of participants in the assessment of research and applications on natural hazards. Washington, D.C.: Joseph Henry Press.

Pudasaini, S. P., & Hutter, K. (2007). *Avalanche dynamics [electronic resource]: Dynamics of rapid flows of dense granular avalanches*. Berlin: Springer.

Quarantelli, E. L., & Kreps, G. A. (1972). *A perspective on disaster planning*. Retrieved from Columbus: Ohi.

Sanoff, H. (2000). *Community participation methods in design and plannning*. John Wiley and Sons.

Schneebeli, M., Laternser, M., Föhn, P., & Amman, W. (1998). *Interactions between climate, avalanches and technical measures (Wechselwirkungen zwischen Klima, Lawinen und technischen Maßnahmen)*. Zurich: vdf Hochschulverlag AG.

Schweizer, J., Bruce Jamieson, J., & Schneebeli, M. (2003). Snow avalanche formation. *Reviews of Geophysics*, *41*(4), 1016. http://dx.doi.org/10.1029/2002rg000123.

The Federal Assembly of the Swiss Confederation. (1991). *Federal law on forest*. Retrieved from. https://www.admin.ch/opc/de/classified-compilation/19910255/.

The Swiss Federal Institute for Snow and Avalanche Research. (n.d.). History of the SLF Retrieved from http://www.slf.ch/ueber/geschichte/index_EN.

Swiss Disaster Management Professional. (2013). *Interview by Maria Kornakova*. Semi-structured interview, Davos.

Switzerland. (1972). *Federal Decree on urgent measures in the area of spatial planning from 17 March 1972 (Bundesbeschluss über dringliche Massnahmen auf dem Gebiete der Raumplanung vom 17 März 1972)* (Vol. 12). Sammlung der Eidgenössischen Gesetze.

Vasilescu, L., Asmatullah, K., & Himayatullah, K. (2008). Disaster management cycle–A theoretical approach. *Management & Marketing-Craiova* (1), 43–50.

Wamsler, C. (2014). *Cities, disaster risk and Adaptation*. London, New York, NY: Routledge.

Wilhelm, C., Wiesinger, T., Bründl, M., & Ammann, W. J. (2001). The avalanche winter 1999 in Switzerland-an overview. In *Paper presented at the Proceedings International snow Science Workshop, Big Sky, Montana, USA, 1–6 October 2000.*

WSL Institute for Snow and Avalanche Research SLF. (n.d.-a). Avalanche protection. Retrieved from http://www.slf.ch/forschung_entwicklung/lawinen/lawinenschutz/index_EN.

WSL Institute for Snow and Avalanche Research SLF. (n.d.-b). Behavior of dense flow avalanches. Retrieved from http://www.slf.ch/forschung_entwicklung/lawinen/fliessverhalten/index_EN.

CHAPTER

8 Reconstruction of Informality: Can Formal Reconstruction Re-create Informality?

Mojgan Taheri Tafti

The University of Tehran, Tehran, Iran

INTRODUCTION

Informality is increasingly seen as a key feature of urban planning and development in many cities of the Global South (McFarlane, 2012; Miraftab, 2012; Roy, 2005). This feature, as argued by Roy (2009b), creates a certain "territorial impossibility of governance, justice and development." In this chapter, I examine this proposition by focusing on the urban reconstruction process in two disaster-affected cities. Urban reconstruction is characterized as the "compression of urban development activities in time" (Olshansky, Hopkins, & Johnson, 2012). Therefore, informality is expected to be entangled in urban reconstruction policies and planning, and to be more conspicuous given the momentum for urban transformation after disasters. I argue that different actors involved in reconstruction activities, including planners and nongovernmental entities, must learn to work with the realities of cities of the Global South, including informality.

Recent scholarship (Alsayyad & Roy, 2006; McFarlane, 2012; Miraftab, 2012; Roy, 2012) theorizes urban informality beyond traditional conceptions of unregulated economic activities or settlements in cities (McFarlane, 2012). Instead, informality is understood as continuous and arbitrary shifts between what is formal/informal or legal/illegal within a continuum of formal and informal. In this sense, urbanization is taking place in a context where the law is often "rendered open-ended and subject to multiple interpretations and interests" (Roy, 2005). This state of deregulation and arbitrary decisions creates the impossibility for urban planning and governance in the sense of achieving the intended outcomes.

Looking at the two earthquake-affected cities—Bhuj in India and Bam in Iran—this chapter seeks to address the understudied question of how informality played a role in the transformation of urban environments during the reconstruction phase. My goal is not to evaluate postdisaster reconstruction policies and planning in these two cities but to highlight some of the distinctive challenges

Urban Planning for Disaster Recovery. http://dx.doi.org/10.1016/B978-0-12-804276-2.00008-6

and paradoxes that informality presents for "building back better." In particular, I focus on a common theme that emerged from these two case studies: the dispossession and displacement of marginalized groups from well-located urban land and the accompanying production of new or perpetuated landscapes of risk following urban disasters. I explain who was displaced and how, and which legal, extralegal, or market-based mechanisms are to be challenged by disaster grassroots organizations, planners, and independent humanitarian actors for averting such displacements. These mechanisms, I will argue, can be challenged by a strategic use of informality and adopting a rights-based agenda, which highlights the role of location in realizing human rights in cities.

The chapter draws primarily on field research I conducted in 2010–12 in Bhuj and Bam to investigate their long-term recovery after the earthquakes of 2001 and 2003, respectively (Tafti, 2015). The selection of the two cases was based on similarities in the nature of the disasters, population size, the relative contemporaneity of the two events, similarities in housing reconstruction policies, and the role of the state as the major actor in urban reconstruction. My intention is not to compare and contrast the reconstruction in these two cities but rather to undertake a cross-case analysis where the experiences of one context can deepen our understanding of the other. I draw on 20 interviews with the chief planner and senior bureaucrats in central or state governments and heads of departments who are directly responsible for postdisaster reconstruction, examining policy documents and internal or published reports of the relevant institutions (in particular, the Housing Foundation of Iran (HFIR) for the case of Bam, and the Area Development Authority (BHADA) and Gujarat State Disaster Management Authority (GSDMA) in the case of Bhuj), and 95 semistructured interviews with disaster-affected people in both cities.

The rest of the chapter is structured as follows: It first presents a brief review of the recent debates on informality in cities of the Global South. Disaster and reconstruction efforts in the context of the two case studies are explained, and the ways in which urban informality played a role in the displacement of marginalized groups and in shaping or reshaping landscapes of dispossession and risk after the disaster are assessed. The chapter concludes with a discussion of possible ways to restore the rights of marginalized groups through a strategic use of informality.

INFORMALITY AND THE PRODUCTION OF URBAN INEQUALITY

The "Southern turn" in urban studies (Rao, 2006) has opened up new theoretical avenues for understanding and interpreting urbanization (Robinson, 2006; Watson, 2009). Integrating diverse experiences of urbanization from cities of the Global South to mainstream urban studies has not only provoked more reflection on, or reassessment of, existing perspectives and theories but has also highlighted the necessity of developing a greater range of theoretical frameworks and starting points to interpret processes of urbanization in diverse contexts (Parnell & Robinson, 2012).

Urban informality, as one of these starting points, is increasingly considered as a key feature in urban planning practices and scholarship (Miraftab, 2012; Mukhija & Loukaitou-Sideris, 2015). The resurgence of informality in urban studies has been accompanied by a new conceptualization of this notion. Earlier accounts of informality were mainly concerned with the deep inequality in access to urban infrastructure and services within the cities of Global South (Kudva, 2009). In particular, they focused on two dimensions of informality (albeit mainly separately): first, entrepreneurial activities and creative ways that people devise to survive and move forward; second, the spatial manifestation of informality that emerges through self-help housing backed by sweat equity and incremental consolidation (Ghertner, 2014), taking place on urban land, often without formal, registered tenure. Urban informality, in this sense, has often been considered as a response to structural problems such as poverty, and the inability of the state and market to respond to the needs of an increasingly urbanized population to adequate housing, jobs, and services. These responses are often characterized as flexible, pragmatic, calculated, and autonomous actions and as attempts to redefine and renegotiate urban space (Bayat, 2007; Kamel, 2014).

Recent theorizations of urban informality have gone beyond a focus on these autonomous activities of survival and their spatial manifestations (McFarlane, 2012). Informality, instead, is seen within a formal–informal continuum, where formality and informality are not fixed and are subject to the arbitrary decisions of different actors, including the state (McFarlane, 2012; Roy, 2009b). The arbitrary interpretations of what is formal or informal, legal or illegal makes the law open-ended (McFarlane, 2012; Roy, 2009b). As a result, the use, purpose of, or access to urban infrastructure, resources, and land can hardly be "fixed and mapped according to any prescribed set of regulations or the law" (Roy, 2009b). This state of deregulation presents a challenge for urban planning and governance in terms of achieving intended outcomes.

This more recent perspective locates urban informality not only in the domain of the urban poor but also with that of the elites and the state. Urban informality, in the form of unfixed and ambiguous practices of naming, managing, governing, and producing urban development, is seen as an integral part of the territorial practices of the powerful segments of the society (Roy, 2009b). Informality in this sense lays down "the rules of the game, determining the nature of transactions between individuals and institutions and within institutions" (Alsayyad & Roy, 2006). In fact, underlying the issue of informality are differences in access to economic and political resources (Mukhija & Loukaitou-Sideris, 2015). The state in particular, by placing itself outside the law and making arbitrary decisions on what is formal and informal, plays a key role in creating a "particular form of elite urban development," and maintaining an unequal urban development (Ong, 1999). As noted by Altrock (2012), the task before us is then "to track the different ways in which informality and formality are put to work as resource, disposition, practice, or classification in the production of urban inequalities."

BHUJ EARTHQUAKE AND RECONSTRUCTION

On January 26, 2001, Gujarat State in India was struck by an earthquake measuring 7.7 on the Richter scale. In Bhuj (Fig. 8.1), a city with a population of around 130,000, approximately 7000 people died. Most casualties were living in high-density, old urban fabric—a walled city—where 50% of the buildings were destroyed (Balachandran, 2005). Around 11,036 houses collapsed and 27,617 were partially damaged. Before the earthquake, 40% of the population were tenants, mostly residing in the old urban fabric (Burns & Tiwari, 2008).

The State Government of Gujarat formulated a recovery program, funded by two loans of US $771 million from the World Bank and the Asian Development Bank. For Bhuj, the newly prepared urban development plan suggested a combination of in situ

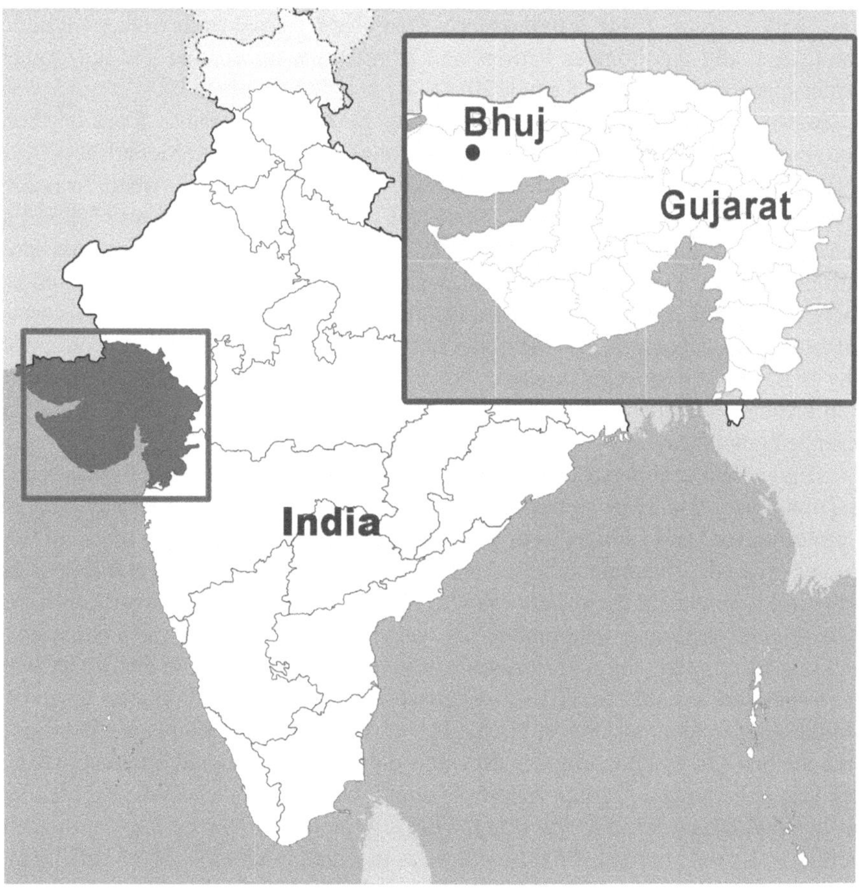

FIGURE 8.1

Location of Bhuj in Gujarat, India.

building and relocation for the reconstruction of the city. The state government allocated a disproportionate amount of the reconstruction funds to the reconstruction of Bhuj and modernization of its urban infrastructure, maintaining that Bhuj would set an example for future urban projects in the state.[1] Pursuing this modernization project, a land pooling and readjustment mechanism was used to reorganize 12,000 small and irregular-shaped parcels of land in the walled city to provide land for roads and amenities (Ballaney, 2008). Furthermore, the new development plan restricted the height of buildings and the permissible floor space index. These changes implied a horizontal expansion of the city. Three relocation sites were accordingly developed outside the walled city on government-owned land to accommodate approximately 5500 households. Relocation to these sites was voluntary, but the allocation of 100 m^2 plots—bigger than most plots in the walled city—at desirable locations encouraged relocation.

Policies of financial assistance distribution to households were mainly concerned with housing reconstruction for homeowners. The owner-occupiers of destroyed houses (6402 homeowners) could receive a maximum of US $3225 for building a 45 m^2 house. Renters could receive subsidized land in relocation sites or could wait for the decision of their landlords on the reconstruction of their rental units by state assistance. In 2004, the negotiation of Abhiyan (a regional NGO) and a local community-based organization resulted in introduction of a new policy for low-income renters who could not afford the cost of land and housing construction. Under this policy, 450 households received a house built by Abhiyan in a site known as GIDC (Gujarat Industrial Development Corporation). The site was initially allocated for temporary housing outside the city and later accommodated those without a housing recovery option such as low-income renters and squatters.

BAM EARTHQUAKE AND RECONSTRUCTION

Two years after the Bhuj earthquake, on December 26, 2003, the historic city of Bam (Fig. 8.2) in Iran was affected by an earthquake registering 6.6 on the Richter scale. With a population of 104,469, Bam lost 23,503 people—almost one-fourth of its population. More than 80% of buildings in the city and around 24,598 urban housing units were severely damaged (World Bank, 2010). Before the earthquake, around 18.8% of the people were renters (Ghafory-Ashtiany & Mousavi, 2005).

Postearthquake policy responses were formulated by the central government. Financing the recovery program relied primarily on public funds and a US $220 million loan from the World Bank. The new development plan of Bam suggested an in

[1] In line with what is known as the general bias toward urban areas in Gujarat's overall development strategy, a substantial portion of financial resources and efforts were concentrated in cities and particularly in Bhuj. The reconstruction and renovation of urban infrastructure (in 14 towns and cities) received US $147 million, over one-third of which (US $51 million) was allocated to reconstruction of infrastructure in Bhuj. For more details on the modernization of infrastructure in Bhuj, see http://bhujada.com/galleries/bhada-gallery/.

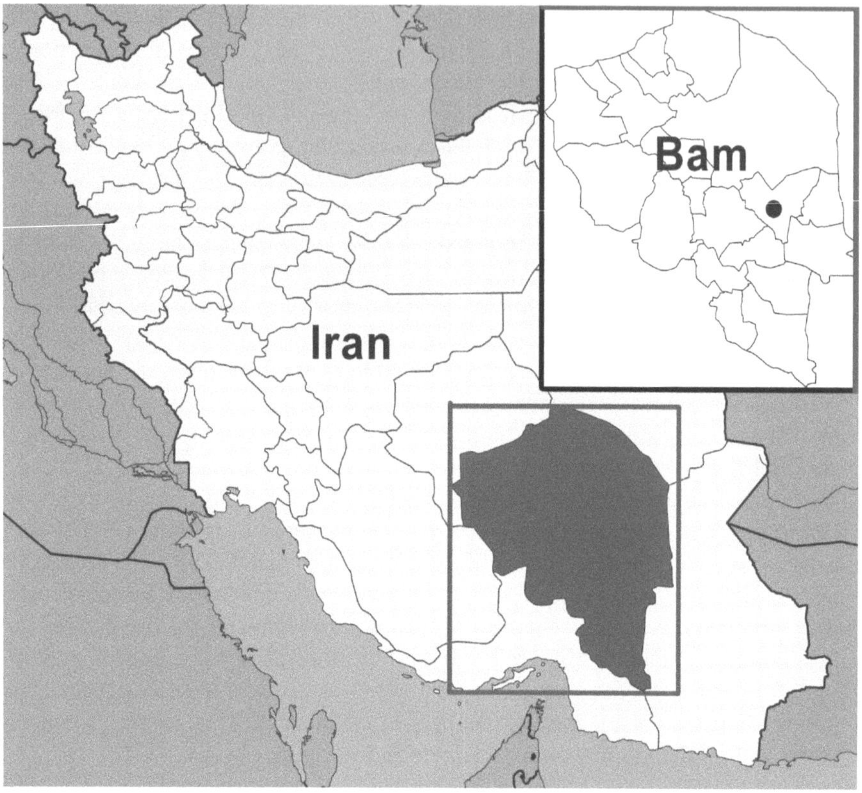

FIGURE 8.2

Location of Bam in Iran.

situ reconstruction for the city. Unlike the recovery programs in Bhuj, which focused on economic growth, in Bam the economic recovery of the city was almost totally overlooked (World Bank, 2010), and officials narrowly interpreted recovery as the reconstruction of damaged buildings. Like Bhuj, policies of assistance distribution to households were mainly concerned with housing reconstruction for homeowners. According to these policies, property owners were eligible to receive a maximum of US $17,647 for building an 80 m^2 house, for each damaged house they owned. Two years after the earthquake, renters—in addition to new couples—became eligible for receiving a grant, provided that they bought a plot or could build a second unit in their extended household's plot.

POLITICAL AND SPATIAL PRACTICES OF INFORMALITY IN RECONSTRUCTION

This section maps the ways in which the mutually constitutive political and spatial practices of informality played a role in the transformation of urban space in Bam and Bhuj after the earthquake. These practices deepened urban segregation and shaped and reshaped landscapes of dispossession in the form of new patterns of informal living in inner urban areas and a new or thickened layer of poverty at the urban periphery. The foregoing examples are not meant to focus on the plight of individual dispossessed households or particular cases of corruption or mismanagement in reconstruction activities. Rather, my intention is to unbundle some key driving forces that resulted in the displacement of marginalized groups and the challenges and complexities of pursuing a "build back better" agenda where political and spatial practices of informality are entangled in planning and reconstruction activities. These forces drive groups such as low-income renters, sharers, squatters, and migrants to reshape or shape new landscapes of dispossession and risk both inside and outside the city.

In both cities, the state-led assistance distribution and the arbitrary decisions regarding what is formal/informal tenure, who is counted as disaster-affected, and who is eligible/ineligible for receiving public assistance were the key driving force for the displacement of low-income renters, sharers, and squatters. Both reconstruction programs initially rejected any direct allocation of assistance to renters, purportedly due to the often informal nature of tenure arrangement practices in these cities. Another reason for excluding the renters and squatters from public assistance was the conceptualizing of disaster impacts solely based on damages incurred to private endowments. Neither the reconstruction program nor the market-driven process of reconstruction could supply affordable housing at least until 10 years after the earthquake. Firstly, low-income landlords were struggling with their own housing reconstruction. Secondly, the reconstruction of rental units was a rational investment only in areas with higher real estate value, where the higher rent could provide a return on investment quickly. Furthermore, in Gujarat and Bhuj, the policy of assisting landlords to rebuild their rental units providing the restitution of the preearthquake renters was practically abandoned. In this city the reconstruction of rental units was further constrained by much lower regime of development rights in the new urban plan. All these factors made the reconstruction of affordable rental units a very slow process. Ten years after the earthquake, in Sonivad, one of the most affected neighborhoods of Bhuj, only one newly built rental unit was found.[2]

In both cities, the later changes in assistance allocation policies made renters eligible to receive assistance provided that they could afford buying a plot in the city. Access to resources, therefore, became the criterion for establishing eligible/ineligible disaster-affected citizens. The latest policy announced in Bhuj, 7 years after the earthquake, acknowledged that "many" renters could not submit their rental receipts or contracts to benefit from the previous policies. Renters, under this policy, were eligible to buy a 50 m^2 plot at the GIDC site. However, at the time of the last field research in 2012, the policy has still not been implemented, ostensibly due to delays in fixing the land price by the local authorities.

The case of informal settlements in Bhuj is the illustrative of the law becoming open-ended and subject to multiple interpretations. Before the earthquake, 42% of the total urban population were living in informal settlements (Environmental Planning

[2] This house and two other small rental rooms that survived the earthquake were the only rental units in Sonivad, which were found with the help of community leaders and neighbors.

Collaborative, 2002). The recovery program considered assistance allocation and a redevelopment project for the squatters whose houses collapsed due to the earthquake. While, according to the urban development plan, most informal settlements located within the walled city were destroyed (Environmental Planning Collaborative, 2002), the state-led damage assessment considered these houses as damaged and not collapsed. As a result, none of these households received assistance.[3]

RECONSTRUCTION OF THE LANDSCAPES OF INFORMALITY IN INNER URBAN AREAS

Having no housing options, low-income renters, sharers, squatters, and migrants who came from surrounding areas for construction jobs shaped new landscapes of dispossession and risk both inside and outside the city. Inside the city, they resided in the cracks and gaps of the formal city. In Bhuj, these households erected tents and makeshift housing on vacant lots belonging to the state or private landowners (Fig. 8.3). Some of them began to incrementally change their *kachcha* (temporary) huts to houses with more permanent materials, albeit without the state-led technical support program for safe construction.

In Bam, the new landscapes of dispossession were observed in the backyards of single-family housing. Given that the average size of plots in Bam was large, homeowners kept their temporary housing in their backyards and rented them to

FIGURE 8.3

Households are living in makeshift houses on government or private lands in Bhuj.

[3] From approximately 70,000 people living in informal settlements, only 25 households, whose houses were pulled down for construction of new roads, received a plot in GIDC site outside the city and housing assistance.

those who could not find a place to live, including preearthquake renters. Real estate agents in the city are still trading these temporary units. People living in the backyards of their relatives' houses have already started building unpermitted second units, using traditional and unsafe construction methods. Furthermore, like Bhuj, vacant plots provided a space for informal living in the city (Fig. 8.4). Unlike Bhuj, however, these households had to pay rent for living in tents or makeshift units in vacant plots, while having inadequate access to basic services such as water. These plots were mainly palm groves, delibcrately left untended with a view to a future land-use change. Residents of these tents and makeshift houses are mainly migrant households who came to the city seeking labor jobs. According to the 2011 census (SCI, 2011), 1112 households live in this condition in different parts of the city. Of this, only 68 people are living alone (migrant workers) and the remainder are families.

RECONSTRUCTION OF THE LANDSCAPES OF INFORMALITY ON URBAN PERIPHERY

Apart from inner urban areas, the urban periphery also saw new landscapes of informality emerge, shaped by renters, sharers, squatters, and migrants. In Bhuj, as noted, a site in the urban periphery, known as the GIDC site, was designated for temporary housing after the earthquake. Preearthquake renters, sharers, squatters, and immigrants moved to this site, bought or rented these temporary units from homeowners, whose housing construction was finished. As mentioned, 450 houses were later built at this site for low-income renters with the help of a regional NGO, while the rest of

FIGURE 8.4

The abandoned palm grove is rented out to migrants in Bam.

renters and squatters remained in rudimentary housing built as temporary dwellings more than a decade ago. These groups have been dispossessed of the diverse forms of urban life in the walled city including access to jobs, services, and their social network in the face of inadequate rental units and rising rents. These households are living in a limbo condition, or "permanent temporariness" (Yiftachel, 2009), waiting for the implementation of the policy introduced in 2008. Given the 7 km distance from this site to the walled city and inadequate social infrastructure, some[4] of the "beneficiaries" sold or informally rented out their new units and moved to squatter settlements located closer to the city and to job opportunities.

In Bam, finding a housing solution in the urban periphery was driven by the policy that allocated assistance to nonlandowners—e.g., renters, shares, or new couples—provided they could buy a plot. Those who could buy a plot outside the city's official boundaries could receive a lower amount of assistance (assistance for housing in rural places). The prohibitively high rent in the city and the lure of receiving public assistance forced these households to look for a foothold where land was more easily available—in an administrative no-man's-land, outside of municipal boundaries. The newly built houses outside the city (as shown in Fig. 8.5) are located next to the earthquake fault zone and do not have adequate access to utilities and services. These houses were built without the state-led technical support program for safe construction.

INFORMALITY AND STRUGGLES FOR URBAN SPACE

In both cities, arbitrary decisions over what is formal/informal—and hence eligible/ineligible—also led to the dispossession of those with lower economic and political power from the diverse forms of street life they seek to maintain for their livelihoods while directing public funds to the most powerful groups. In both cities the old bazaar was the center of this informal politics.

In Bhuj, planning for the reconstruction of the walled city was revised in favor of the major and powerful merchants and traders in the old bazaar. Simpson (2008) notes that this group campaigned successfully to exempt the bazaar from the road-widening prescription of the new plan, a campaign that was backed by many local politicians with the same caste-base or political-base affiliations. As a result, a compromise was made to cut off just half a meter on average from each shop (Simpson, 2008). The whole project was finished in 40 days and inaugurated by the Chief Minister of Gujarat. As Fig. 8.6 shows, partial demolition of the buildings without any structural strengthening provisions left buildings in this part of the city even more vulnerable to a future earthquake.

At the same time, on pursuing the aim of creating a "modern and clean" city, the new zoning regulations in Bhuj considered "unclean activities," like the Mutton Market, a small local market for selling meat, chicken, and fish, to be

[4] In one section of the site with 92 houses, 53% of the beneficiaries rented out their new houses.

FIGURE 8.5

The formation of new landscapes of informality in urban periphery in Bam. Left: Bam in 2004. Right: Bam in 2014 (Google Maps, 2014a).

FIGURE 8.6

Partial demolition of buildings in the bazaar in Bhuj.

relocated outside the walled city. Shopkeepers, however, resisted this decision and rebuilt the market themselves with temporary materials. Another process that displaced and dispossessed the small retailers in the walled city was the inflexibility of assistance policies that impeded the reconstruction of buildings with mixed commercial and residential use. Some retailers built small structures in their previous trading places to continue their businesses. At the same time, new commercial units were built in the city through public–private partnership arrangements, which were unaffordable and were bought by traders who already had shops in other parts of the city.

In the Iranian case, Bam, the reconstruction of the old bazaar and the commercial center of the city was heavily subsidized and commercial units were built in a state-led development project, almost entirely with public funds. Meanwhile shopkeepers used shipping containers that are brought to the city as temporary working places and shops, for trading, some for nearly 10 years after the earthquake (Fig. 8.7). During these 10 years, local authorities attempted to close down these shops using various methods such as cutting off electricity and water, or even forced closure. These struggles ended in 2013 when local authorities removed almost all these shipping containers from the city.

FIGURE 8.7

Shipping containers used as shops. These containers were removed from the city 10 years after the earthquake.

DRIVING FORCES OF THE DISPLACEMENT

What emerges from the two cases of Bam and Bhuj is "a tangled and confused web of informal and formal actions" (McFarlane, 2012), which played a role in shaping a fractured pattern of urban recovery and new landscapes of dispossession. In many cases these landscapes of dispossession overlap with the new geographies of risk in these cities. Research on urban informality deals with unsettling practices of categorizations such as legal/illegal or formal/informal. In both cities, integrating disaster risk reduction measures—which in the case of earthquakes are often expensive, engineering requirements—with reconstruction activities was consolidated by regulatory instruments that linked and justified legality/illegality based on safe/unsafe construction practices. At the same time, the state-led technical assistance did not cover the informal, self-help construction activities. These seemingly rational and apolitical regulatory instruments, on the one hand, became the subject of interpretation and clientelistic practices, such as the case of the old bazaar in Bhuj, and on the other hand, drove the urban poor back to a fragile built environment.

In both Bam and Bhuj the three major contributors to the displacement and dispossession of marginalized groups were assistance distribution programs, planning and building regulations, and land governance. Informality was entangled in these policies, regulations, and practices, exacerbating the condition of the urban poor.

ASSISTANCE DISTRIBUTION PROGRAMS

In both cities, the centrality of housing tenure—in the form of registered ownership—in counting and recognizing urban citizenship, drove lower income renters and squatters into homelessness. In both cities, informality, in the forms of interpreting policies (for instance, in defining who is eligible/ineligible for housing assistance), or suspending policies (such as allocating land for squatters in Bhuj), worsened the consequences of these policies for lower income groups. This conditional defining of citizenship based on property ownership is not limited to the cases of Bam and Bhuj. Reviews of recent disasters and housing assistance programs (McCallin & Scherer, 2015) have highlighted the wholesale neglect of lower income renters, sharers, and squatters as a recurrent and persistent problem. Leaving the housing recovery of low-income renters to market forces directed them to self-help solutions or pushed them to ownership in peripheral locations with inadequate access to jobs and services, leading to a downward spiral of poverty. No technical assistance was offered for the self-help construction activities of these households. In short, the idea of "build back better" was never actualized for these households.

PLANNING AND BUILDING REGULATION

The new urban plans in both Bam and Bhuj largely repeated the common problems of urban planning in the Global South (Watson, 2009). In Bhuj, for instance, under new planning and building regulations, land parcels could be developed under a much lower regime of development rights compared to preearthquake conditions. Examples of such regulations were building setbacks and a limited permissible floor space index, where a high proportion of plots in the walled city were less than 50 m^2 (e.g., 61% of plots in the Khatri Falia area). These measures considerably lowered the residential space in well-located urban areas and representing what Tibaijuka (2006, as cited in Watson, 2009) has criticized as planning that sweeps the poor away.

Furthermore, neither city considered an initiative for securing land and supplying affordable housing for marginalized groups. Interestingly, in the case of Bhuj, the town planning scheme in Gujarat had the regulatory capacity to allocate land for this purpose through its land pooling and readjustment scheme (Ballaney, 2008). This capacity, however, was not utilized in the reconstruction process. Instead, the new town planning scheme ignored the preearthquake presence of those informal settlements located in prime locations in the walled city. The new development plan of the city simply stated that "most of the slums in the walled city have been affected badly due to the earthquake and have suffered a lot of destruction. There are hardly any people living there now" (EPC, 2002, p. 152). The state-led rubble removal process facilitated the expulsion of the squatters from the walled city.

LAND GOVERNANCE

Reconstruction in both the case of Bam and Bhuj was accompanied with land speculation; the local government turned a blind eye to, encouraged or even initiated, land transfers after the disaster. In Bhuj, the new urban plan introduced three new suburbs in prime locations and housing assistance policies encouraged people to move to these suburbs. As a result, in highly affected areas of the walled city the local government, land speculators, and "higher" caste communities become the major landholders. The local authority faced difficulty in selling its plots in these neighborhoods, and land speculators only started to construct new houses around 8 years after the earthquake. As a result, 39% of plots in Sonivad were still empty in 2011. In Bam, land use change and subdivision of palm groves, which covered 70% of the city area, was banned, partly due to pressure from UNESCO, which considered these palm groves as a part of the cultural heritage of the city. In the absence of any mechanism for protecting these groves, those located in inner urban areas were abandoned with a view to the future land use change (Fig. 8.8) or were subdivided and sold out for residential purposes. In both cities and as a result of this speculative urbanism, large portions of well-located urban areas remained without any good use, while the poor were pushed to the urban periphery or to live in the inner city's "gray spaces," where they "are neither integrated nor eliminated" (Yiftachel, 2009).

CHALLENGING THE DISPLACEMENT DRIVEN BY INFORMALITY THROUGH INFORMALITY

Bam and Bhuj were similar in being relatively well-resourced cities after the earthquake. Therefore, the displacement and dispossession of marginalized groups was not a result of inadequate resources but their inequitable distribution. Informality exacerbated and mediated this inequitable distribution. In this final section, I explore the possibilities for averting these displacements and dispossessions in the aftermath of disasters when the state is unwilling or unable to do so. I explore these possibilities in particular by focusing on the role that planners, humanitarian actors, nongovernmental and grassroots organizations can play. In cases of major disasters, the presence of international institutions and the media can arguably open up new opportunities for these groups to hold the state and donors accountable for the dispossession and displacement of the urban poor.

Recent debates in urban studies (McFarlane, 2012; Miraftab, 2012; Roy, 2005; Watson, 2013) seek opportunities to use informality strategically by grassroots movements and their key partnership with other actors within or outside governmental organizations. Watson (2013) posits that informality need not always be negative; it can be strategically utilized to frame policies to counteract the dispossession and displacement of the poorest of the poor from well-located areas in cities. What is missing from these debates, however, is a clear direction for such practices. Urban informality often—as was the case in Bam and Bhuj—takes the form of interpreting, extending, and suspending policies and regulations. This makes its transformation difficult (Gilbert & De Jong, 2015). I argue that a rights-based agenda would offer a clear direction for challenging extralegal measures that result in the displacement of the urban poor and would place an empowering agenda on the table for holding the state and donors accountable. A rights-based agenda also fits well with the nature

FIGURE 8.8

Loss of palm groves in Mahd-e-ab 8 years after the earthquake (right) compared with the condition before the earthquake (left) (Google Maps, 2014b).

of many humanitarian agencies and is already embedded in their discourses, policies, and practices (Cornwall & Celestine, 2004). Adopting a rights-based agenda, however, requires a clear articulation of rights in the way that it reflects the urban context and the settlement-based expressions of the denial of human rights. Parnell and Pieterse (2010) offer a rights-based urban development framework, which is grounded in an understanding of the locational and settlement-based determinants of the realization of rights. They define these rights as the second and third generation rights[5], with the second generation being linked to household-based rights, such as housing or access to water, and the third generation being settlement-based entitlements, such as access to social amenities and infrastructure. This framework captures the often-overlooked role of location- and settlement-based factors in the impoverishment of the urban poor and offers a moral platform for sustained political pressure and a strategic use of informality from progressive interest groups.

The subsequent paragraphs reflect on the application of three methods that use informality in counteracting the dispossession and displacement of the poor after disasters. While these methods can by no means transform the inequitable distribution of resources after disasters, they represent a series of attempts to advance a rights-based agenda in disaster reconstruction activities.

POLITICS OF INCLUSION

Politics of inclusion (Roy, 2009a) has attracted attention in field of planning and development studies, albeit with different terminologies such as "civic governmentality" or "deep democracy" (Appadurai, 2001). This burgeoning concept in urban and development studies is inspired by the work of organizations such as Shack/Slum Dwellers International in utilizing tools for producing knowledge, including self-enumeration, self-mapping, and self-documentation, by the marginalized groups. This generation of knowledge can present a platform for partnership with state agencies (Appadurai, 2001) and more importantly can be instrumental in advancing the rights-based agenda.

Official interviewees in both Bam and Bhuj noted that renters, sharers, and squatters were overlooked in assistance allocation programs primarily because of the often undocumented nature of their tenure. Squatters account for one-third of the urban population in cities of the Global South (UN-Habitat, 2007); in Bhuj the proportion is 40% of the urban population (Mukherji, 2008). Furthermore, tenure arrangements for low-cost rental units in these cities are often oral and undocumented. Under this circumstance, requesting formal tenure documents for counting and recognizing disaster-affected population implies a wholesale neglect of the poorest of the poor (and many more without sufficient documentation). One way of making these groups visible and less likely to be overlooked is by producing knowledge about them. This can be done even through informal mechanisms such as the testimony of neighbors and mutual identification (Appadurai, 2001). Humanitarian agencies are well placed to initiate, assist, and facilitate these processes as a part of their damage assessment operations. This can enable the poor to negotiate support and access to different

[5] First generation rights are the individual rights to health, education, etc.

external sources including technical assistance. There are limited reported precedents in utilizing this approach in the disaster reconstruction context. In Bachhau, India, a city affected by the 2001 Gujarat earthquake, a local NGO was involved with the community enumeration and mapping for squatters. Based on this information and advocacy, the authorities agreed to regularize plots and support retrofitting of their buildings (McCallin & Scherer, 2015).

INFORMAL POLITICS

Miraftab (2012) highlights the importance of informal politics and innovative practices through which "subordinate groups renegotiate their social spatial relations." The case of Bhuj presented one example of such informal politics. As noted, the negotiation of a group of different organizations with the local authority secured land, basic services, and housing assistance for low-income renters, albeit at the urban periphery and for a limited number of renters. Similar instances were reported in Turkey and Indonesia. After the 1999 Marmara earthquake in Turkey, a cooperative was formed for renters and squatters to procure land for housing projects by government loans. It was, however, only in 2011 and after a court decision, that they were granted land with infrastructure (Arslan & Johnson, 2010). Likewise, in Aceh (Indonesia) after the 2004 tsunami, it was advocacy and lobbying by NGOs that led to the allocation of cash assistance and land to renters and squatters 2 years after the disaster (Steinberg, 2007). However, despite these initiatives securing land and housing for renters and squatters, they could not preclude the displacement of these groups from well-located urban areas.

These examples highlight the necessity of adopting a rights-based urban development agenda in informal politics. Humanitarian agencies have shown an inadequate knowledge on the complexities of urban vulnerability when it comes to land-related issues (Pantuliano & Elhawary, 2009) and the role of access to jobs and services in the urban poor's day-to-day life. As a result, their responses to urban displacement have failed to address the issues of land rights and access to jobs and services in a systematic way (de Waal, 2009). The idea of "shelter" as a human right, without considering land and access in urban areas, has long been challenged by the urban poor with their feet by moving from poorly located donor or state-built housing in the periphery to the areas closer to urban centers.

SCALE JUMPING

Roy (2005) introduces the idea of scale jumping as "a strategic engagement with and utilising the resources and tools available at the global scales." She uses the example of Narmada Dam in Central India, financed by the World Bank. In this case, activists pressured the World Bank, rather than focusing solely on the Indian government, for accountability toward resettlements standards. This process led to changes in the World Bank's safeguards in its next projects. Such negotiations, she argues, indicate "the possibility of pursuing issues that are stymied and silenced at the local level" (p.155).

In disaster reconstruction processes, such as those in Bam and Bhuj, where policymaking is articulated at the national or state level, outside pressure might be viewed as an intrusion. A closer look at these policies in Bam and Bhuj, however, demonstrates that they were, in fact, highly influenced by the flow of knowledge and funding through transnational networks of donors and the humanitarian community. One example of such influence is the adoption of the owner-driven model of housing reconstruction in both cases of Bam and Bhuj, a model that is strongly promoted by the World Bank (Tafti & Tomlinson, 2015). The power of a rights-based discourse can highlight the responsibility of the donors for the negative impacts on people flowing from projects they funded. This is particularly the case when the World Bank, as the major funding source for reconstruction in major disasters in the Global South including Bam and Bhuj, views itself as being "in a position to influence post-disaster reconstruction policies" in these countries (Jha, Barenstein, Phelps, Pittet, & Sena, 2010).

International institutions, however, have made slow progress in innovating and adapting their responses to the particularities of urban environments (Pantuliano, Metcalfe, Haysom, & Davey, 2012). While the World Bank's loan safeguards show considerations for issues such as development-induced involuntary relocations or the importance of location in rural areas, they have not yet grasped the role of location and settlements in realization of rights in cities (Parnell & Pieterse, 2010). Another problem with scale jumping is that international organizations and donors are often inert in making changes in general and in particular in cases of politically sensitive issues. For instance, in Bam the international entities were aware that women without land rights[6] did not have any alternative for housing recovery. The UNDP (2008) briefly noted that the government asked them to contribute to the housing reconstruction of female-headed households only if they owned a plot in the city, and as a result, the UNDP excluded nonlandowners from its program.

CONCLUSION

In this chapter, I examined the uneven geography of urban reconstruction in two cities—Bhuj and Bam—and highlighted some of the distinctive challenges and paradoxes for "building back better" in cities of the Global South. In these two case studies, a focus on urban informality, as McFarlane (2012) suggests, served as a form of urban critique in that it sought to expose the ways that claims of what is formal or informal mediate the access of different social groups to resources including aid, urban land, and services. Such critique, however, cannot present the whole picture. The informality perspective highlights how different interpretations or suspensions of policies added complexities and challenges for achieving the intended outcomes, but it is less engaged with the question of whether those intended outcomes are desirable and for whom. In Bam and Bhuj, assistance

[6] Widows, according to the civil laws of the country, could not inherit land from their deceased spouses.

distribution policies, urban planning, and land governance were mostly tailored to the benefit of the more powerful segments of the population, and the informality entangled in these policies, regulations, and practices exacerbated the condition of the poor. The result was the dispossession and displacement of renters, sharers, and squatters from well-located urban areas and formation or re-formation of landscapes of informality and risk inside the city and on the urban periphery.

Also in practice, informality is difficult to challenge, given that it often takes the forms of interpreting, extending, or suspending policies and regulations. Researchers recently discussed how informality, in the forms of politics of inclusion, informal politics, and scale jumping, can be strategically utilized in the struggles of the urban poor. I argued that in adopting such strategies, there is a need for a mobilizing agenda and suggested that a rights-based approach, which recognizes the role of location in realizing human rights in urban areas, can offer a common ground for humanitarian actors. Avoiding and minimizing the involuntary relocation of property owners in reconstruction or development projects has already been included in the safeguards of international organizations such as the World Bank. Extending this idea to minimize the displacement of people without a formal ownership title, from where they live or work, however, remains a political endeavor that involves planners, humanitarian, and grassroots organizations.

ACKNOWLEDGMENT

I would like to thank Prof. Carolyn Whitzman and Prof. Richard Tomlinson for their constructive comments on the first draft of this chapter. This research is partly funded by Melbourne Social Equity Institute at the University of Melbourne.

REFERENCES

Alsayyad, N., & Roy, A. (2006). Medieval modernity: On citizenship and urbanism in a global era. *Space & Polity*, *10*, 1–20.

Altrock, U. (2012). Conceptualising informality: Some thoughts on the way towards generalisation. In C. McFarlane, & M. Waibel (Eds.), *Urban informalities: Reflections on the formal and informal* (pp. 171–194). Surrey: Ashgate.

Appadurai, A. (2001). Deep democracy: Urban governmentality and the horizon of politics. *Environment and Urbanization*, *13*, 23–43.

Arslan, H., & Johnson, C. (2010). Turkey: Can small actors overcome the absence of state will? In M. Lyons, T. Schilderman, & C. Boano (Eds.), *Building back better: Delivering people-centred housing reconstruction at scale* (pp. 263–284). Warwickshire, UK: Practical Action Publishing Ltd, Schumacher Centre for Technology and Development.

Balachandran, B. R. (2005). *The reconstruction of Bhuj case study: Integration of disaster mitigation into planning and financing urban infrastructure after an earthquake*. The World Bank.

Ballaney, S. (2008). *The town planning mechanism in Gujarat, India*. Washington, DC: The World Bank.

Bayat, A. (2007). Radical religion and the habitus of the dispossessed: Does Islamic militancy have an urban ecology? *International Journal of Urban and Regional Research*, *31*, 579–590.
Burns, T., & Tiwari, D. (2008). *Post-disaster land issues: Case study of the 2001 earthquake in Gujarat, India, post-disaster land case studies*. Global Land Tool Network.
Cornwall, A., & Celestine, N.-M. (2004). Putting the 'Rights-Based approach' to development into perspective. *Third World Quarterly*, *25*, 1415–1437.
Environmental Planning Collaborative. (2002). *Development plan of Bhuj the government of Gujarat, urban development and urban housing department, Gandhinagar, Gujarat, India*.
Ghafory-Ashtiany, M., & Mousavi, R. (2005). History, geography, and economy of Bam. *Earthquake Spectra*, *21*, 3–11.
Ghertner, D. A. (2014). India's urban revolution: Geographies of displacement beyond gentrification. *Environment and Planning A*, *46*, 1554–1571.
Gilbert, L., & De Jong, F. (2015). Entanglements of periphery and informality in Mexico city. *International Journal of Urban and Regional Research*, *39*, 518–532.
Google Earth. (2014a). Bam (2003 and 2014), [map], 29°06'21.49" N 58°22'55.28" E, Eye alt 1039 m. *Retrieved from*. http://www.earth.google.com.
Google Earth. (2014b). Bam (2003 and 2014), [map], 29°06'08.88" N 58°21'20.78" E, Eye alt 1039 m. *Retrieved from*. http://www.earth.google.com.
Jha, A. K., Barenstein, J. D., Phelps, P. M., Pittet, D., & Sena, S. (2010). *Safer homes, stronger communities: A handbook for reconstructing after natural disasters*. Washington, DC: The World Bank.
Kamel, N. (2014). Clandestine urbanization: Reconstituting urban space in the margins of the Phoenix metropolitan area. *Journal of Urbanism: International Research on Placemaking and Urban Sustainability*, 1–12.
Kudva, N. (2009). The everyday and the episodic: The spatial and political impacts of urban informality. *Environment and Planning A*, *41*, 1614.
McCallin, B., & Scherer, I. (2015). *Urban informal settlers displaced by disasters: Challenges to housing responses*. International Displacement Monitoring Centre.
McFarlane, C. (2012). Rethinking informality: Politics, crisis, and the city. *Planning Theory & Practice*, *13*, 89–108.
Miraftab, F. (2012). Planning and citizenship. In R. Crane, & R. Weber (Eds.), *The Oxford handbook of urban planning* (pp. 787–802). New York: Oxford University Press.
Mukherji, A. (2008). *Negotiating housing recovery: Why some communities recovered while others struggled to rebuild in post-earthquake urban Kutch, India*. Berkeley: Department of Architecture. University of California.
Mukhija, V., & Loukaitou-Sideris, A. (2015). Reading the informal city: Why and how to deepen planners' understanding of informality. *Journal of Planning Education and Research*. http://dx.doi.org/10.1177/0739456X15591586.
Olshansky, B. B., Hopkins, L. D., & Johnson, L. A. (2012). Disaster and recovery: Processes compressed in time. *Natural Hazards Review*, *13*(3).
Ong, A. (1999). *Flexible citizenship: The cultural logics of transnationality*. Durham: Duke University Press.
Pantuliano, S., & Elhawary, S. (2009). *Uncharted territory: Land, conflict and humanitarian action*. London: Humanitarian Policy Group; Overseas Development Institute.
Pantuliano, S., Metcalfe, V., Haysom, S., & Davey, E. (2012). Urban vulnerability and displacement: A review of current issues. *Disasters*, *36*, S1–S22.
Parnell, S., & Pieterse, E. (2010). The 'right to the city': Institutional imperatives of a developmental state. *International Journal of Urban and Regional Research*, *34*, 146–162.

Parnell, S., & Robinson, J. (2012). (Re) theorizing cities from the Global South: Looking beyond neoliberalism. *Urban Geography*, *33*, 593–617.
Rao, V. (2006). Slum as theory: The South/Asian city and globalization. *International Journal of Urban and Regional Research*, *30*, 225–232.
Robinson, J. (2006). *Ordinary cities: Between modernity and development*. London, New York: Routledge.
Roy, A. (2005). Urban informality: Toward an Epistemology of planning. *Journal of the American Planning Association*, *71*, 147–158.
Roy, A. (2009a). Civic governmentality: The politics of inclusion in Beirut and Mumbai. *Antipode*, *41*, 159–179.
Roy, A. (2009b). Why India cannot plan its cities: Informality, insurgence and the idiom of urbanization. *Planning Theory*, *8*, 76–87.
Roy, A. (2012). Urban informality: The production of space and practice of planning. In R. Crane, & R. Weber (Eds.), *The Oxford handbook of urban planning* (pp. 691–705). New York: Oxford University Press.
SCI. (2011). *National population and housing census*. Statistical Centre of Iran – Provincial Data – Kerman Province.
Simpson, E. (2008). *The rise and fall of collective public action in the aftermath of the Gujarat Earthquake of 2001*. NGPA working paper series. London: Non-Governmental Active Programme, Department of Social Policy, London School of Economics and Political Science.
Steinberg, F. (2007). Housing reconstruction and rehabilitation in Aceh and Niaz, Indonesia-rebuilding lives. *Habitat International*, *31*, 150–166.
Tafti, M. T., & Tomlinson, R. (2015). Best practice post-disaster housing and livelihood recovery interventions: Winners and losers. *International Development Planning Review*, *37*, 165–185.
Tafti, M. T. (2015). Housing assistance distribution after disasters: Does it enable affected households to recover? *Environmental Hazards*, *14*, 361–377.
UNDP. (2008). *Final review: Sustainable housing reconstruction programme in Bam through community mobilization and participation*. Iran: UNDP Country Office.
UN-Habitat. (2007). Slum dwellers to double by 2030. Nairobi. *Retrieved from*. http://www.preventionweb.net/files/1713_463146759GC202120Slum20dwellers20to20double.pdf.
de Waal, A. (2009). Why humanitarian organizations need to tackle land issues. In S. Pantuliano (Ed.), *Uncharted Territory: Land, conflict and humanitarian action* (pp. 9–26). Rugby: Practical Action Publishing.
Watson, V. (2009). 'The planned city sweeps the poor away…': Urban planning and 21st century urbanisation. *Progress in Planning*, *72*, 151–193.
Watson, V. (2013). Planning and the 'stubborn realities' of global south-east cities: Some emerging ideas. *Planning Theory*, *12*, 81–100.
World Bank. (2010). *Implementation completion and results report, Bam earthquake emergency reconstruction project*. The World Bank.
Yiftachel, O. (2009). Critical theory and 'gray space': Mobilization of the colonized. *City*, *13*, 246–263.

CHAPTER

Reconstructing Vulnerability After the 2013 European Floods: Oil Damage and Recovery

9

Mark Kammerbauer[1], Christine Wamsler[2]

[1]Nuremberg Institute of Technology, Nuremberg, Germany; [2]Lund University Centre for Sustainability Studies (LUCSUS), Lund, Sweden

INTRODUCTION

An increase in disasters around the world has been observed in the past decades (UNISDR, 2015). In the Indian Ocean, in New Orleans, in Haiti, in Japan, on the east coast of the United States, and, in 2013, in Europe, there have been disasters caused by tidal waves or floods that have led to the destruction of housing and infrastructure, environmental pollution, evacuation, waves of refugees and death (Munich Re, 2015a, 2015b).

In the aftermath of these disasters, the same questions arise (cf. Gaillard et al., 2014; Kammerbauer, 2013; Rodriguez, Quarantelli, & Dynes, 2007; Sipe & Vella, 2014): why does the affected population not succeed in quickly reconstructing and returning to their homes? Or may it actually be appropriate to return to places that are at high risk at all? We take the areas of Fischerdorf and Natternberg in the Lower Bavarian city of Deggendorf, Germany, which were strongly affected by the 2013 European floods, as a case to address these questions.

We assess how local and regional emergency management and urban planning institutions, volunteer initiatives, and the impacted population have contributed to and influenced the reconstruction and inherent adaptation processes. What implications did and do the 2013 floods have for them, was risk reduction achieved, and does vulnerability play a role here?

The assessment is based on a literature review, interviews with 10 key informants, a survey,[1] participatory observation,[2] and site visits carried out (independently) by

[1] A quantitative questionnaire survey with a sample group of 55 individuals was held in 2014. It made use of the snowball method to identify potential participants (Flick, 2012). Questions asked built upon the insight gained through the interviews with key informants and dealt with potential vulnerabilities, degree of damages of housing, insurance, and adaptation measures undertaken.

[2] Participant observation took place during two public events: the "Symposium Stadt Land Flut" of the BYAK (Bayerische Architektenkammer = Bavarian Chamber of Architects) on September 25, 2013, at the "Haus der Architektur" in Munich and the "Baufachtag Dingolfing," the Building Expert Meeting of the LVS Bayern (Landesverband der Sachverständigen Bau = State Association of Expert Surveyors in Construction) on November 15, 2013, at the City Hall in Dingolfing, Bavaria.

Urban Planning for Disaster Recovery. http://dx.doi.org/10.1016/B978-0-12-804276-2.00009-8

the authors during 2013–2014 to gain an understanding of how the affected communities have rebuilt and adapted after the disaster impacts. At this point in time, the recovery process is still ongoing. Many former residents were either still in the process of rebuilding their homes or had not yet returned. Possible trauma needed to be considered to address ethical concerns of field research (O'Mathuna, 2010). Purposeful sampling was used to select the interviewees. They were approached according to their activity as representatives of particular groups, members of key organizations, or experts of a particular field of action (Flick, 2012). The written notes and transcripts served as a basis for a qualitative content analysis (Flick, 2012).

THE 2013 FLOODS IN EUROPE

In May and June 2013, a steady downpour with rainfall levels up to 300% above the monthly norm led to extremely high water levels in Germany, Czech Republic, and Austria. Germany's Elbe and Danube rivers caused massive damage across the country. Six eastern and southern German states were hit especially hard. An emergency was declared in a total of eight federal states of Germany and 56 municipalities, leading to evacuations of 80,000 people from their homes. According to the German Insurance Association, there were 180,000 insurance claims, totaling 2 billion euros. The German Ministry of the Interior estimated damage totaling nearly 1.5 billion euros to private households and residential buildings in the federal states affected (Federal Ministry of the Interior, 2013).

As a result of the floods and the damage caused, the German Federal Government and the federal states entered into agreements on immediate assistance to implement measures in support of private households. These were to be supplemented in the medium term and also in the long term by financial grants for reconstruction. Aid for the reconstruction of damaged buildings has been administered and paid by the federal states. This is jointly financed by the federal and state governments under the Reconstruction Aid Act and through a special fund totaling 8 billion euros. The portion borne by the Federal Government for private households and residential buildings is almost 600 million euros (Federal Ministry of the Interior, 2013).

THE CASE OF DEGGENDORF: FISCHERDORF AND NATTERNBERG AREAS

After a dike failed, water levels rose in Deggendorf to a record-breaking 8 m. The town flooded, and in Fischerdorf, Natternberg, and the abbey village of Niederalteich, buildings were 2 m underwater, with oil-contaminated water lapping at second story windows. In Fischerdorf, it took 6 days for the peak level to be reached, stretching over an area of 7 × 6 km. The town of Deggendorf was no longer accessible from the motorways, as these were also flooded and had to be closed to traffic. Five thousand people in this area were asked to voluntarily evacuate their homes. Some evacuees (with working social networks) went to stay with relatives and acquaintances, whereas others found temporary accommodation in emergency shelters (for example, civic centers or sports centers in the region). Those evacuated had to leave their pets and possessions, including items of sentimental value, photographs, and valuables, behind. There was also temporary accommodation for helpers from the Federal Armed Forces and the Federal Police. The crisis response was coordinated at the Deggendorf federal police barracks, and attempts were made to repair the breach in the embankment at Fischerdorf as quickly as possible (Kallus, 2013a, 2013b, 2013c, 2013d).

The scale of the disaster surpassed the fears of everyone involved, and it became apparent that the flood resulting from the breaching of the embankment was not the only problem. While the water was able to flow away in Niederaltaich because of the sloping topography at that location, the floodwater in Fischerdorf and Natternberg could not be pumped away until 11 days had passed. Some houses in these areas were submerged up to their roof levels. Household fittings were consequently damaged beyond repair.

To make things worse, the floodwater was mixed with heating oil and petrol from burst tanks and with animal carcasses and waste, similar to the "toxic gumbo" that caused great distress in New Orleans after the devastating Hurricane Katrina. In the case of Deggendorf, heating oil tanks had been ripped away from their anchors in residential buildings, so that the oil was spilled and spread over the surface of the water. As of now, this toxic layer can still be seen on shrubs and trees. In addition, because of the long period that the flooding lasted, heating oil penetrated the materials of the surrounding buildings. The strong odor of oil is also still perceptible in some houses. Nearly 1000 houses were affected in the wider district of Deggendorf, 600 of them in Fischerdorf and another 90 in the settlement of Natternberg. Particularly in Natternberg Siedlung, houses were predominantly built in the 1950s; the typical porous cinder block used in construction at that time is particularly susceptible to oil intrusion.

EMERGENCY ASSISTANCE

The people affected have received assistance not only from state institutions but also from nongovernmental organizations (cf. Kammerbauer & Wamsler, 2017; Wamsler, 2016). Considerable numbers of volunteers were deployed to assist the residents on the scene and to collect donations, including the "Freunde durch Helfen" ("Friends Through Help") campaign of the Straubinger Tagblatt/Landshuter Zeitung group of newspapers, the "Deggendorf räumt auf" ("Deggendorf Cleans Up") campaign organized by students or the citizen-led "Deggendorf Hilft" group ("Deggendorf Helps"). In addition, the Malteser relief agency staff met at the chapel in Fischerdorf every week for coordination purposes. Social media were used in particular to organize volunteer helpers and invite donations. Nonprofit organizations and charitable associations collected donations totaling more than 100 million euros, which also included financial and material compensation for wrecked household effects.

The clear-up of apartments and houses began in the second week of June 2013. Staff from the Fire Brigade, the Federal Agency for Technical Relief, and Bavarian Red Cross helped to pump away the water and clear up the debris. This work was done house by house, including removing dirt and oil. In this context, homeowners were also given assistance by the many volunteers.

The spilled heating oil covered floors, walls, and ceilings on the inside and outside with a thick, slimy coat. It took 6 weeks until oil and dirt had been removed from surfaces. As a result of the damage, complete renovation was often necessary, and buildings were possibly at risk of collapse. One thousand homes in the district had been affected, and at least 500 in the Fischerdorf and Natternberg areas were impacted (Kallus, 2013a). In addition, because of the extent of heating oil pollution and the depth to which the oil–water mixture penetrated into the affected building structure, more houses than originally hoped for were deemed inhabitable and had to be or would have to be demolished. Upon the request of homeowners, cases had to be assessed individually by expert surveyors.

RECONSTRUCTION

Regarding the financing of the reconstruction, the federal and state governments agreed to pay 80% of the reconstruction costs incurred by homeowners without insurance coverage resulting from flood damage. This financial support is provided for a period of 3 years. Money from donations was used to contribute to meeting the remaining costs and was coordinated by a local charitable donation board. In addition, there is also a hardship fund to allow complete support to be provided to those in need, managed by the administrative district office (Landratsamt Deggendorf), a regional agency that is also responsible for emergency management.

The administrative district office, the city authority of Deggendorf, and the Rural Development Office (ALE) as representative of the Free State of Bavaria coordinated their planning activities to assist those affected during reconstruction. The Rural Development Office had been involved in a project on the development of settlements in Fischerdorf, Natternberg, and Altholz before the flooding occurred. The administrative district office, as the authority responsible for disaster management, has 15 administrative officers. After the disaster, some of them have specialized in particular consequences of flooding, such as demolition of residential buildings, household effects, etc. Demolition approvals were issued and planning applications were made as early as July 2013. In addition, the city authority's building department (Stadtbauamt) was involved in surveying building reconstruction and supporting required adaptation measures.

The widespread oil pollution made the reconstruction process particularly difficult. In Fischerdorf and Natternberg, all but one of the residential buildings was heated with oil. Because of the spatial distribution of oil within the flooded area, it was not only the buildings with oil tanks that were affected but the spreading of heating oil combined with the floodwater also led to pollution in surrounding buildings.

A meeting of experts was held in Deggendorf to discuss the oil contamination in July 2013. In addition, the rural development office organized consultation and a preliminary inspection of the buildings affected, for example, to establish their structural stability. In the case of renovation, specific measures to address oil pollution are necessary, irrespective of where oil pollution occurred. Experts were appointed to advise those affected after the floods. The town of Deggendorf shared in the funding of this advisory activity, and related technical cooperation with the administrative district office began at the end of July. However, there were major problems in making personnel and time available to deal with the preparation of expert assessments, which the homeowners paid for. In addition, testing samples as part of an expert assessment can take several months.

There was a widespread need to remove layers of plaster contaminated with heating oil in Fischerdorf. Initial decontamination measures were rapidly taken internally and externally by some of those affected. In this initial action, plaster was stripped and floor, wall, and ceiling coverings were removed. However, in many cases, brown stain patches appeared through the newly applied plasterboard or plaster, which indicated persistent heating oil pollution. Quick responses taken by residents thus proved in many cases to be their disadvantage, causing additional personal and financial distress and frustration.

Professional testing of hydrocarbon content based on test drillings, which started later, was supposed to show whether pollution exceeded values posing a danger to health. If this was the case, people's entire work with clearing of debris and mud, cleaning, partial demolition, decontamination, and reconstruction had been in vain. Demolition is then the last remaining option, if possible followed by new construction on the same plot of land.

A specific condition set by the federal state to be met for new construction measures after the disaster was that replacement buildings were to be of equivalent kind and use, and 75% of the former basement area could be added to the living area. Thereby, the construction of a basement, which would contribute to the risk of flooding, should be avoided. In addition, sleeping rooms were forbidden at ground floor levels. The administrative district office as the responsible authority on site for disaster management had to ensure that the principle of equivalence is put into practice, while the city authority dealt with the authorization of the building applications.

After demolition, questions arose regarding the disposal of contaminated rubble. Refuse separation is necessary here. Demolition material polluted with heating oil is recyclable, but the costs are five times higher than for untreated building rubble. In the case of a detached house, this means that 60,000 euros must be spent on treating the building rubble. On-site visits provided evidence that this worst case scenario affected many residents. In many places, the new buildings erected after the floods already had a roof structure by the end of the year 2013. By spring of 2014, the first houses had already been completed. However, there were many vacant buildings, such as large apartment blocks or plots of land already cleared, alongside which there were heaps of demolition material. According to the regional district administration, at least 25 families who had rented their apartments had been known to have left the region.

VULNERABILITY

The assessment of the emergency and reconstruction processes has revealed different vulnerability factors, which have translated into people's (high or low) capacity to recover and adapt to future risk,[3] such as:

- time availability and capacity (e.g., language skills) to deal with administrative procedures,
- social networks (friends and family) for temporary accommodation and support,
- health conditions necessary to actively engage in the recovery process,
- access to insurance and/or state compensation,
- financial assets (cash) available for rebuilding purposes,
- forms of living (being house owners or tenants),
- contacts/linkages with construction firms or other (help) organizations, and
- linkages with governmental and nongovernmental organizations and associated power structures.

The process of application for reconstruction support was sluggish among eligible uninsured homeowners. There was a great uncertainty over the future among those affected. Rumors and false information began to spread regarding who would receive funding. Residents were irritated by funding decisions, e.g., if their neighbors received assistance: "why did they get a lawn mower and we didn't?" In certain cases there were problems in submitting applications. As a result, in some cases no applications at all were made, although funding would have been possible. The application process was complicated, to such an extent that people were not able to fill out forms without help. In particular, elderly people and people with different ethnic backgrounds than the majority German population were initially quite passive in searching for assistance or even rejected financial help. Filling in applications by themselves proved problematic for many senior citizens or citizens with language difficulties. As described by one individual: "Especially elderly or people with migration backgrounds did not know how to access recovery assistance ... how to fill in the forms" (cf. Wamsler, 2016). Helpers (both from the city authority and volunteers) therefore offered support, in particular, to this group of people. In addition, practical assistance was given in everyday life, for example, driving elderly people to the shops or photocopying documents for financial support application processes. Beyond that, many of the people affected received help from family members, whereas others complained about an absence of family support. Furthermore, there was consequential harm in the form of physical and mental illnesses (including suicides), both caused directly from the disaster impacts and indirectly through the recovery process. Several residents therefore abandoned the goal of reconstruction and stayed in contaminated houses or abandoned the wish to return and moved away.

[3] Similar factors have been discussed in literature by Birkmann (2006), Bolin (2007), Bürkner (2010), Cannon (2008), Cannon and Müller-Mahn (2010), Nelson, Adger, and Brown (2007), Oliver-Smith and Hoffman (2002), and White and O'Hare (2014).

However, most people affected did not make any complaints at public events, which can be seen as a sign of their resignation or mistrust in the offered assistance.

Not everyone affected was insured since some areas were not covered by insurance, or insurance policies were too costly for the inhabitants. However, it turned out that there were a surprisingly large number of people insured. At 30%, Bavaria has a relatively low degree of insured homeowners compared to other federal states, such as Baden-Württemberg. Having insurance meant that people could not apply for financial compensation from the public purse and had to wait for the insurance to pay, a process that generally took much longer. Consequently, having insurance resulted in slower recovery, leading to widespread frustration and conflicts between those affected. Insurers appeared to prefer renovation to demolition and rebuilding, so that there were borderline cases due to the contamination with heating oil, and there were cases where the insurance payouts were not sufficient to renovate a building.

Furthermore, there was a shortage of building and construction firms for reconstruction and, consequently, people were competing for their services. The costs of tradespeople and building activities rose. As described by a resident:

> *The waiting times presented a real hindrance in people's efforts to recover ... since processing damage analyses and compensation was slow. The analyses of the oil contamination and damages took several months... People did not know if they needed to demolish their houses or not, if they could renovate, ... and if yes, how ... And after several months of waiting, all construction firms were already contracted by others, and there were hardly any skilled labour available... This led to increased tension between all sides and stress to citizens, until today (July 2014).*
>
> **cf. Wamsler (2016).**

It was surprisingly found that even building firms from Passau came to Deggendorf, a town that was also badly affected by flooding. People also mentioned that assistance was not always distributed equally, especially when staff of voluntary organizations also included some of the affected population.

The problems in reconstruction faced by the owners of buildings also had consequences for tenants. Temporary alternative accommodation was needed for tenants during reconstruction. Tenants moved away and found new places to rent in the region. However, rental costs still rose after the flood. In addition, the problem emerged as to where those who decided to remain in Fischerdorf would move to. They mostly relocated within the rural district of Deggendorf, for example, and locations at higher altitudes were chosen because of the experience of flooding, while other properties were left empty.

A "virtual" tour through Fischerdorf can illustrate the differentiated vulnerability of the reconstruction process. There is an apartment block where a tenant on the ground floor has already moved away. One person is still living on the first floor of another block but intends to move to stay with his sister. The two residents of a neighboring detached house are affected by oil leaking from another building. Both are senior citizens, one of the couple being in need of care. Two neighboring families from an ethnic minority are extremely grateful for the support given by volunteer helpers in submitting an application for a reconstruction grant, although it took them some time to gain confidence in the process. An elderly gentleman is living on the first floor of another detached house. The basement and the ceiling above are affected by oil pollution. Tests for possible heating oil penetration were performed with three test drillings per storey. The outcome was that parts of the outside wall of the basement and the ceiling above the basement room had to be demolished. While the demolition work is in progress, the owner is living on the upper floor of the house while being exposed to potential health risks due to oil contamination. Finally, there are some residents who are fully engaged in the recovery process, with little time for other duties, being both affected and volunteering for (or employed by) different organizations, including governmental and nongovernmental organizations.

The described differences in vulnerability translate to different levels of capacity to deal with hardship, financial and health impacts, administrative matters, and the ability to cooperate and engage in city–citizen collaboration for recovery and more long-term adaptation but have not been taken into account by the assistance offered.

PLANNING STRATEGIES

Planning for recovery after disaster comprises a complex field of social and spatial interrelations where an interest in betterment of existing conditions is key to reducing risk, enhancing resilience, and enabling climate change mitigation and adaptation—yet, results may be uneven (Filion, Sands, & Skidmore, 2015; Olshansky & Chang, 2009; Pelling, 2003; Vale & Campanella, 2005; Wamsler, 2014). Also in Deggendorf, attempts were made to reduce risk and to deal with the recovery problems faced. However, these did not address the identified vulnerabilities described above. There were calls for creating "green" or ecological compensation areas for new construction, developing renaturation projects, and stipulating landscapes as nature protection areas. In addition, natural gas was to be used for heating instead of oil. Reference was made to the "Flächensparendes Bauen" ("Land-Saving Construction") alliance, with a call for greater densification and avoidance of extensive urban sprawl or large car parks for commercial estates. New urban developments and construction were seen increasingly problematic due to the resultant reduced capacity for drainage. Another result was the rural development program "Flussaue" ("Floodplain") to upgrade the Danube waterway, and for this a sum of nearly 3.5 million euros was made available by the Bavarian State until 2017.

In general, "better" urban planning was demanded, without much specification of what this entails. There were calls for new building typologies or temporary uses that reflect risk reduction or adaptive urban planning measures in architectural terms. However, mechanisms to deal with the historic built heritage and related preservation issues were not discussed. Villages like Niederaltaich, which reflect a long history of settling in these riverine landscapes, are almost entirely located within the floodplain. Some experts were more specific with their suggestions: a rethinking process is considered necessary for building components and materials, as floods occur repeatedly. To support this process, the Bavarian Association of Experts ("Bayerischer Landesverband der Sachverständigen") is planning to formulate rules to identify and remove heating oil damage due to flooding.

In addition, a new resolution adopted by professors of hydraulic engineering and engineering hydrology at German universities from June 2013 (Schumann et al., 2013) was concerned with the requirements to be met by structural flood protection measures in Germany. It was pointed out that settlements along rivers could not be left without engineered flood protection structures, even if they could not be fully implemented. An appeal was made to those affected to take responsibility themselves. At the same time, a professional debate was initiated: "What risks are we willing to take in the future?" In terms of technical measures, a combination of improved embankment structures and controllable polders or retention areas for floodwater was aimed for. However, this seems only appropriate transnationally, in turn necessitating parallel related governance structures. However, many residents objected to these plans. Farmers and municipalities in the Oberpfalz area formed a protest alliance with the head of the district authority and the district assembly. There was criticism that, as a result of designating polder embankments for flood protection, there would be a threat to the drinking water supply and to stable flood protection. In addition, it was feared that affected plots of land would decline in value, against the background of an increased demand for scarce land.

CONCLUSIONS

As the example of Deggendorf, Fischerdorf, and Natternberg shows, characteristics, such as age, ethnic minority status, state of health, and questions as to who is the owner of the building and who is the tenant, who receives funding or has good contacts and who does not, are empirically of great significance to how individual reconstruction proceeds. This inevitably becomes an issue of urban planning and inherent adaptation processes, which is, however, not yet addressed in practice. In the case of Deggendorf this issue was to some extent identified and addressed through cooperation between local and regional actors and nongovernmental organizations. Yet, an active participation of citizens is seemingly absent. The institutional and policy landscape as well as operational interventions to mitigate, respond to, and recover from hazards generally do not involve mutual support through collaborative arrangements between cities and their citizens.

Although vulnerability plays a major role in reconstruction, it is rarely discussed in official documents. It does not appear in documents on flood risk management in Bavaria (State Ministry of the Environment, 2013a) and is equally absent in the floods handbook of the federal government (Federal Ministry of Transport, Building and Urban Development, 2013), in the report on the June 2013 floods issued by the Bavarian State Office for the Environment (State Ministry of the Environment, 2013b) or in the cabinet report on the 2013 floods (Federal Ministry of the Interior, 2013). It is necessary that in future more attention will be paid to this in dealing with the planning aspects of the consequences of disasters, as intended by the European Union in disaster risk management and urban planning in the context of adaptation mainstreaming (European Environment Agency, 2012; Wamsler, 2014, 2016).

Vulnerability as an important factor in urban planning is gaining in importance, when an increase in the intensity and frequency of hazards and disasters necessitates combining goals of reconstruction with aspects of adaptation. It was shown that institutional assistance for recovery can obstruct, discourage, or support citizens' efforts to recover and adapt to future risk, which involves vulnerability reduction. Whether rapid disasters or slow climate impacts—the way in which societies and cities deal with risk is essentially informed by vulnerability in a world of dwindling resources, even in developed industrial nations.

REFERENCES

Birkmann, J. (Ed.). (2006). *Measuring vulnerability to natural hazards*. Tokyo, New York, Paris: Towards Disaster Resilient Societies.

Bolin, B. (2007). Race, class, ethnicity, and disaster vulnerability. In H. Rodriguez, E. L. Quarantelli, & R. R. Dynes (Eds.), *Handbook of disaster research* (pp. 113–129).

Bürkner, H.-J. (2010). *Vulnerabilität und Resilienz. Forschungsstand und sozialwissenschaftliche Untersuchungsperspektiven*. IRS: Working Paper 43.

Cannon, T. (2008). Vulnerability, "innocent" disasters and the imperative of cultural understanding. *Disaster Prevention and Management, 17*(3), 350–357.

Cannon, T., & Müller-Mahn, D. (2010). Vulnerability, resilience and development discourses in context of climate change. *Natural Hazards, 55*(3), 621–635.

European Environment Agency. (2012). *Urban adaptation to climate change in Europe. Challenges and opportunities for cities together with supportive national and European policies*. EEA. Report 2.

Federal Ministry of the Interior (BMI). (2013). *Bericht zur Flutkatastrophe 2013: Katastrophenhilfe, Entschädigung, Wiederaufbau*. Bundesministerium des Innern.

Federal Ministry of Transport, Building and Urban Development (BMVBS). (2013). *Hochwasserschutzfibel. Objektschutz und bauliche Vorsorge* (5. Auflage). Berlin: Bundesministerium für Verkehr, Bau und Stadtentwicklung.

Filion, P., Sands, G., & Skidmore, M. (Eds.). (2015). *Cities at risk. Planning for and recovering from natural disasters*. Burlington, Surrey: Ashgate.

Flick, U. (2012). *Qualitative Sozialforschung. Eine Einführung*. Hamburg: Rowohlt.

Gaillard, J. C., Glantz, M., Kelman, I., Wisner, B., Delica-Willison, Z., & Keim, M. (2014). Taking the 'naturalness' out of natural disaster (again). *Natural Hazards Observer, XXXVIII*(3) 1, 14.

Kallus, P. (June 15, 2013a). *Medienrummel in den versunkenen Siedlungen*. Donau-Anzeiger.

Kallus, P. (June 15, 2013b). *Größte Katastrophe seit dem Zweiten Weltkrieg*. Donau-Anzeiger.

Kallus, P. (June 12, 2013c). *Vor den Trümmern der Existenz*. Donau-Anzeiger.

Kallus, P. (June 10, 2013d). *Das große Aufräumen hat begonnen*. Donau-Anzeiger.

Kammerbauer, M. (2013). Schismo-urbanism: Cities, natural disaster, and urban sociology. *Disasters Journal*, *37*(3), 401–419. http://dx.doi.org/10.1111/disa.12005.

Kammerbauer, M., & Wamsler, C. (2017). Challenging the Consensus: Marginalization Effects in Post-Disaster Recovery (forthcoming).

Munich Re. (2015a). *NatCatSERVICE. Loss events worldwide 1980–2014. 10 costliest events ordered by overall losses*. München: Münchener Rückversicherungs-Gesellschaft.

Munich Re. (2015b). *NatCatSERVICE. Loss events worldwide 1980–2014. 10 deadliest events*. München: Münchener Rückversicherungs-Gesellschaft.

Nelson, D., Adger, N., & Brown, K. (2007). Adaptation to environmental change: Contributions of a resilience framework. *Annual Review of Environment and Resources*, *32*, 395–419 398.

O'Mathuna, D. P. (2010). Conducting research in the aftermath of disasters: Ethical considerations. *Journal of Evidence-Based Medicine JEBM*, *3*, 65–75.

Oliver-Smith, A., & Hoffman, S. (2002). *Catastrophe & culture: The anthropology of disaster*. Santa Fe: School of American Research Press.

Olshansky, R., & Chang, S. (2009). Planning for disaster recovery: Emerging research needs and challenges. *Progress in Planning*, *72*, 195–250.

Pelling, M. (2003). *The vulnerability of cities*. London: Natural Disasters and Social Resilience.

Rodriguez, H., Quarantelli, E. L., & Dynes, R. (Eds.). (2007). *Handbook of disaster research. Handbooks of sociology and social research*. New York: Springer.

Schumann, A., Dittrich, A., Fröhle, P., Graw, K.-U., Grünewald, U., Haberlandt, U., … & Wieprecht, S. (2013). *Versagt der Hochwasserschutz in Deutschland? Resolution von Professorinnen und Professoren der Lehrstühle für Wasserbau und Ingenieurhydrologie an deutschen Universitäten vom 15.06.2013*.

Sipe, N., & Vella, K. (2014). Relocating a flood affected community: Good planning or good politics? *Journal of the American Planning Association*, *80*(4), 400–412.

State Ministry of the Environment (LfU). (2013a). *Hochwasserrisikomanagement. Hochwasserrisikomanagement-pläne*. Augsburg: Bayerisches Landesamt für Umwelt.

State Ministry of the Environment (LfU). (2013b). *Junihochwasser 2013. Wasserwirtschaftlicher Bericht*. Augsburg: Bayerisches Landesamt für Umwelt.

UNISDR. (2015). *Sendai framework for disaster risk reduction 2015–2030*. UNISDR.

Vale, L., & Campanella, T. (2005). *The resilient city: How modern cities recover from disaster*. Oxford.

Wamsler, C. (2014). *Cities, disaster risk, and adaptation, routledge series on critical introduction to urbanism and the city*. London: Routledge.

Wamsler, C. (2016). *From risk governance to city-citizen collaboration: Capitalizing on individual adaptation to climate change*. Environmental Policy and Governance.

White, I., & O'Hare, P. (2014). From rhetoric to reality: Which resilience, why resilience, and whose resilience in spatial planning? *Environment and Planning C: Government and Policy*, *32* 1–17:2.

CHAPTER

The Opportunity for Improved Regulations After the 2009 Victorian Wildfires[1] in Australia

10

Maria Kornakova[1,2], Alan March[1]

[1]*The University of Melbourne, Melbourne, VIC, Australia;* [2]*Massey University, Palmerston North, New Zealand*

INTRODUCTION

The "usual" practice of the postdisaster recovery is unfortunately rebuilding in a way similar to predisaster conditions. Lately, however, more nations and communities are actively using the so-called "window of opportunity" timeframe immediately postdisaster events to implement changes in disaster prevention and risk reduction practices, including those relevant to urban planning. While it can be argued that one event is not sufficient for significant changes in decision-making, there is some evidence of gradual implementation or "buildup" of a scientific base for new regulations and the use of a disaster as a catalytic event to implement them. Using a case study of the post-2009 Victorian Bushfire season, this chapter provides an overview of relevant changes to the bushfire planning practices in the state and focuses on roles of political voices in the decision-making.

Commencing with a discussion of wildfires in Victoria, this chapter sets out the context for wildfire risk management being increasingly oriented toward urban planning and building controls over time. A description of building codes first introduced in 1991 and their eventual integration over time into urban planning mechanisms is provided. It describes the changes in these fields and raises number of political, community, and agency-related issues associated with them, particularly the opportunity for relatively rapid change. This chapter goes on to describe the interactions of state and local politics, and the interactions between these processes with technical and planning processes. To begin, the chapter commences with a description of wildfires in Victoria, Australia.

[1] In Australian settings *bushfire* is used to refer to *wildfire*; thus these terms are used interchangeably in this chapter.

Urban Planning for Disaster Recovery. http://dx.doi.org/10.1016/B978-0-12-804276-2.00010-4

WILDFIRES IN VICTORIA, AUSTRALIA

The Australian Charcoal Database demonstrates the influence of fire over a long period, beginning even from when the continent itself was first formed. Bushfires have been part of the country's ecosystem for millennia, suggesting that the country and its ecosystems have been shaped in large part by fire (Bradstock, Williams, & Gill, 2012). Thus, the likelihood of fire events in most vegetated parts of the country is a question of "when" rather than "if." More recently, the climate of Australia has become more extreme: the frequency of droughts has been increasing and cities are facing severe water constraints (Hennessy et al., 2007). Despite the fact that other natural disasters, such as floods, may take more human lives, bushfires in some regions of Australia are one of the most devastating hazards, with one of the highest economic impacts on communities in several states with the potential to destroy towns and settlements, and surrounding wildlife (Gangemi, Phillips, Stewart, Martin, & Marton, 2003; Hennessy et al., 2007; Packham, 1992). Located in the southeast of the country (Fig. 10.1), Victoria is one of the most wildfire-prone areas in Australia, and indeed the world (Australian Emergency Management Institute, 2006). Wildfires in this area are annual events, which often have significant impacts on humans, their

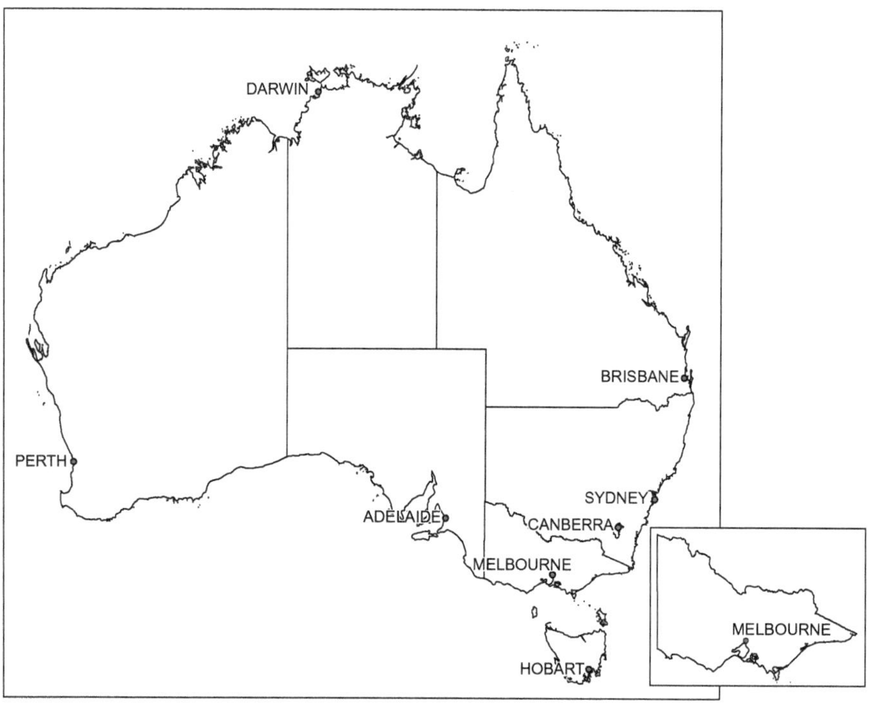

FIGURE 10.1

Map of Australia.

assets, and the environment. Over the past 110 years, for example, there have been 260 recorded bushfire events that have resulted in 825 civilian and firefighter fatalities (Blanchi et al., 2014).

BEFORE THE EVENT: THE WILDFIRE MANAGEMENT OVERLAY

For a country with such a rich fire history and significant impacts on human settlements and assets (Blanchi et al., 2014), formalized urban planning and building policies and requirements that deal with wildfire are relatively new practices, which partially explain some of their shortcomings. In 1991 Australian Standards introduced building standards, AS 3959, for construction in the bushfire prone areas (BPAs), which specifies requirements for timber-based constructions, oriented to dwellings. These standards were optional at the time of introduction and were developed by a group of science and building professionals, with the Commonwealth Scientific and Industrial Research Organisation (CSIRO) as a leading research group included in both development and adoption of the standard. Working solely on the scientific side allowed professionals here to be separated from political and interest group aspects of this problem and to seek proper applications of science to these building and planning policies. This approach generally allows the CSIRO to stay neutral in regard to policy reform processes. Since 1991, there have been two updated versions of AS 3959, and the third undertaken after the devastating events of the 2009 Victorian Bushfire season.

This building standard became mandatory in 1994, when some bushfire risk areas were mapped, requiring all new development within the delineated areas to comply with AS 3959. These areas were named BPAs. According to a bushfire building professional interviewed, the BPA is:

> *a building regulation matter. So if you are in a bushfire prone area you need to then design your building to suit the particular area you're in. So depending on what the risk is. So back in [19]94 when it was first introduced, there was a strange standard and if you were in the bushfire prone area you had to design to the standard.*
>
> **Professional 5 (2015)**

The process of the BPA development is explained by a building professional:

> *The latest information we had at the time and all the science and all the details and all of the considerations had all been put in there. And that standard will continue to be updated. So we introduced that. So any new home built to the standard can now comply with the new standard. And the new standard requires you to assess the site if you're in the bushfire prone area, if you're* not *in the bushfire prone area don't do anything. Bushfire prone area you assess the site, establish where your bushfire attack level is, know what the level of risk, and then you apply that on the standard. The standard, whatever is the level, applies the construction requirements from the standard to your design. That is how you build a house.*
>
> **Professional 5 (2015)**

The BPA itself makes no provisions or requirements for landscaping or planning permits for structures (Professional 2, 2015). Moreover, the only planning regulation at the time, which partly addressed bushfire, was the Special Building Overlay (SBO). The SBO was simply a mapping-based trigger to assess a number of potential disasters and was considered somewhat ambiguous by council planners. Accordingly, to address these concerns, bushfire risks were later addressed in the Wildfire Management Overlay (WMO), which replaced the SBO in October 1997 under Clause 44.06 (Gibson, Carew, Dwyer, Jerome, & Mitchell, 1997).

The policy basis of the WMO aimed to ensure development achieved fire protection objectives that development did not increase potential threats to human lives and assets. Fulfillment of these aims was ensured by the introduction of various fire protection requirements, e.g., sufficient water supply, access to emergency vehicles, design of structure, vegetation controls, and defendable space, which serves as an area for defending the building envelope from fire front, allowing to retrieve the structure when needed (Victoria Planning Provisions, 1997). While the WMO was also map based as a "trigger," it integrated building standard approaches with planning and urban design approaches on the basis of individual site assessments.

Implementation of the WMO was individual for each municipality in consultation with Country Fire Authority (CFA) professionals. Local councils were required to accept all conditions and objectives suggested by CFA professionals. Therefore, the guidelines provided by CFA became crucial for application and provided more details on decision-making tools for bushfire risk assessment (Victorian Bushfires Royal Commission, 2010c). This approach allowed the tailoring of mapping based on the level of risks in the given area, but it also meant that the process itself was rather costly and time-consuming; by 2009 only 35 of 82 local authorities had the WMO in place, and most of these were not extensive areas.

The WMO was applied to progressively larger mapped areas in the time period between 1997 and 2011, after which it was updated and renamed as the Bushfire Management Overlay (BMO). In the period 2006–09 the number of permit applications referred to the CFA was 2,866, with only 24 receiving objections, 88 accepted without change, and 2754 approved with permit conditions (Victorian Bushfires Royal Commission, 2010c).

THE 2009 FIRES

In 2009 the southeast region of Australia suffered the most devastating bushfire disasters among those recorded. The total amount of individual fire events reported to the CFA during the bushfire season of that year is 39,987 (Victorian Bushfires Royal Commission, 2010a) with the deadliest and most severe events occurring on February 7, 2009. Also referred to as "Black Saturday," on that single day fires directly affected 78 communities, burnt a total of 365,020 ha of land, killed 173 people, and destroyed 2056 houses (Country Fire Authority, 2012; Victorian Bushfires Royal Commission, 2010b).

Seeking to understand the causes of such severe impacts and to develop recommendations for new practices, policies, and mechanisms to minimize risks in unavoidable future disaster events, the Australian government established the Victorian Bushfire Royal Commission (VBRC). The investigation determined that on one side of the disaster, there was a "natural" cause of fuel high levels of formation (a prolonged and severe drought in period 1999–2009, lack of sufficient rainfall, and a prolonged heat wave in the last week of January 2009) and natural causes of ignition[2] and vast spread of fires across the state on the day, although man-made assets were acknowledged to exacerbate fire intensity (Victorian Bushfires Royal Commission, 2010a).

In terms of the human and built environment aspects of the 2009 fires, detailed investigations revealed that urban planning and building controls applied in the state before these events were inconsistent, potentially affecting risk levels in individual communities. For example, the most severely affected townships of Kinglake and Marysville did not have the WMO in place at the time of the disaster, and most of the structures in these communities were well below modern standards in terms of bushfire risk management. As a result of the investigation, 19 of 67 recommendations were made regarding planning and building standards and were aimed toward increased consistency in mapping exercises, risk levels, and application of the policies and mechanisms across the state. Two standards that were directly challenged by the VBRC were the WMO including its accuracy and application, and building standard AS 3959.

Investigations identified a number of flaws in WMO policy and suggested revision of the overlay and policy to address them. More particularly:

> *For areas mapped before 2002 the WMO was applied more restrictively [meaning smaller spatial areas] than BPAs and was applied only to areas where controlling a high-intensity fire would be difficult, rather than to all areas where bushfire was likely to pose a threat to life and property;*
>
> *Since July 2002 the criteria for mapping the WMO have been the same as those used for BPAs, but there has been no systematic re-examination of the WMO mapping completed before that time;*
>
> *DSE [Department of Sustainability and Environment] initially opposed applying the WMO to public land – a matter that was not resolved for some time. In December 2005 the Minister for Planning approved application of the WMO to public land, but DSE notified only councils that were in the process of amending their planning schemes at the time, so this criterion was not applied to all planning schemes.*
>
> **Victorian Bushfires Royal Commission (2010c, p. 219)**

[2] With the exception of arson cases reported in the media investigations (Farnsworth, 2012; Russell, 2013).

Other identified flaws included a lack of clear bushfire objectives with a particular focus on management of native vegetation. Moreover, the WMO failed to acknowledge the availability of areas where the safety of people requires clearing land around dwellings. The WMO did not adequately allow site-tailored risk treatments and did not recognize different bushfire risk levels and behaviors in different areas. It was also suggested that water supply requirements specified for the WMO be revised due to the failure of reticulated systems in Marysville (Victorian Bushfires Royal Commission, 2010c, p. 219). A lack of planning requirements for bushfire bunkers was also considered another flaw of policy because important features such as siting, defendable space, access, and egress to such bunkers are not specified. It was also recommended that there be a review of permit triggers and exemptions in relation to whether current uses that require a permit are appropriate and whether small-scale alterations to the dwelling should require permit (Victorian Bushfires Royal Commission, 2010c).

Inconsistency among municipalities regarding the use of the WMO and its relatively limited application pre-2009 is reflective of typical implementation processes in the state, which requires a series of steps. The most basic of these are mapping, ground truthing, public consultations, public panels to hear any objections, approval to proceed, council decisions, and ministerial approval. Time, costs, and unpredictability of application outcomes of the WMO in planning processes were proposed to be eliminated by introducing a single bushfire hazard mapping tool and apply it to the entire state once mapping is finished (Victorian Bushfires Royal Commission, 2010c).

Despite the flaws of the control, the WMO was nonetheless a good demonstration of a shift from using solely building code fire prevention practices in Victoria, to a more comprehensive and integrated approach, which included planning controls. While there have been some problems identified with application of the WMO, the fire events of the 2009 Victorian Bushfire season demonstrated that houses where the WMO was applied had a significantly lower rate of loss when compared to those that were not (Holland, March, Yu, & Jenkins, 2012). The WMO included slope, topography, and native vegetation in its assessment criteria, demonstrating the application of evidence or scientific base in planning mechanisms. AS 3959 provides additional information on the flammability of materials used for construction in BPAs. Research conducted shortly after the events of the 2009 Victorian Bushfire season also demonstrated that the WMO appeared to have had positive impacts on bushfire protection of houses. Loss of houses was significantly lower in areas that had undergone risk treatment as per WMO requirements, as compared to those that were not assessed and treated under earlier planning and building regulations. Data were collected in five communities: Kilmore East—Murrindindi; Churchill—Jeeralang, Delburn, Beechworth, and Bunyip. The total number of dwellings both within WMO and outside its boundaries is 4288, of which 1632 (38.06%) were destroyed during 2009 Victorian Bushfire season. Furthermore, the number of dwellings within the fire area and WMO was 1412, of which 569 (40.30%) were destroyed (Holland et al., 2012). These data demonstrated the potential of the WMO as a planning tool to reduce risks associated with fire disasters, while also demonstrating that the WMO, on its own, is not capable of providing absolute protection of dwellings; hence, additional prevention mechanisms are required.

CENTRALIZATION AND WIDESPREAD APPLICATION OF THE WMO

The effectiveness of the WMO was questioned by the VBRC authorities after the 2009 bushfire season, and this response has been widespread across the state. As a result, more rigorous and accurate mapping was developed and the WMO was updated including a new overlay introduced on May 18, 2011. The BMO is a set of planning regulations that trigger the need for planning permits in areas with increased risks for human lives from bushfire disasters (Victoria Planning Provisions, 2013).

The BMO, in contrast to the WMO, was applied in a way that resulted in a more simplified and "one-size-fits-all" manner. This resulted in some areas being deemed as highly risky but allowed for quicker, arguably less expensive and consistent applications across the state. While some might see it as a band-aid solution, the overall mapping exercise was rigorous and the tool is considered more effective compared with the WMO (Professional 2, 2015). Another difference between the two were the stated objectives. Compared to the WMO, the BMO is both more precise and also includes broader urban planning goals, such as assisting in strengthening community resilience to bushfire and identification of hazardous areas requiring special measures; ensuring location, construction and design of development considers implementation of bushfire protection measures; and ensuring priority of safety to both human life and property by reducing bushfire risks to an acceptable level (Victoria Planning Provisions, 2011).

According to the interviewed bushfire planning professional, despite some differences in "a nutshell":

> *WMO to BMO is pretty much the same thing. That's all it is. It's a trigger for planning. After Black Saturday they [the state government] declared the whole of Victoria in the Bushfire Management Overlay. So if you were in Melbourne, CBD, you were in the BMO. That was just a knee jerk reaction to government.*
>
> **Professional 2 (2015)**

The similarity between the two tools supports the argument that the BMO remains an integrated tool of planning and building controls. However, the separate processes of BPA and BMO mapping also demonstrate a level of disparity between planning and building institutions as there is still no single map that can provide a more holistic approach and a set of solutions that BMO delivers. As stated by the interviewed bushfire planning professional, there is a need to ensure that there should be:

> *just one map for all Victoria, [...] and there should be the bushfire planning, [there] should be building, [there] should be prescribed burning or burning up, [there] should be community, and all of that should be one map. That's the ideal, which is not happening right now*[3]*. Which is what you've got is they're all in complete isolation.*
>
> **Professional 2 (2015)**

[3] Comment made in May 2015.

This statement demonstrates the disparity of all institutions involved in the "ideal" of integrated practice, which would include both planning and building controls. It further highlights the need to find ways to integrate these relevant institutions. The disparity between these is partially addressed when the changing roles of the agencies are analyzed, as below.

THE BMO AND REALLOCATION OF ROLES

After the 2009 fires, bushfire planning in Victoria was in a state of flux, not just in terms of regulation but also in terms of roles and institutions. A shortage of experts with knowledge in both planning and bushfire engineering science led to the employment of fire responders as the key consulting experts in the post-2009 planning assessment processes. This was later considered inappropriate, when many parties expressed the opinion that the agency was "not fit for purpose" as noted by a bushfire manager interviewed (Professional 2, 2015). Established in 1850 (CFA, 2012) the CFA is not a policy writing or advice agency, but rather they are agents "whose job is to fight fires, who have a deliberately risk-averse structure…we would never under any other circumstance go to the fire brigade and say 'how fire safe do you want the community to be?' because their answer will always be 'as safe as we can imagine'" (Professional 4, 2015). Therefore, it is not only "not fair to ask the fire brigade to put risk of a bushfire in a broader policy framework" (Professional 4, 2015) but it also resulted in the development of policy that was a somewhat extreme execution of the precautionary principle, possibly breached the constitutional rights of residents and ultimately led to a backlash against the policy (Professional 2, 2015; Professional 4, 2015; Professional 6, 2015). Under the new BMO regulations, the referral authority for planning site assessments was the CFA. This resulted in a many ongoing conflicts, as the CFA's main goal is to ensure that properties and lives are minimized, and they have rather conservative views on the risk assessment. In contrast, planners and other professionals involved sought to balance out risks with many other facets informing decision-making.

THE CHALLENGES OF APPLYING SCIENCE

Another issue associated with the new regulations was the lack of a sufficient science base for the standards applied. As discussed above, the BMO by and large is a precautionary tool, which lacks more detailed and tailored risk assessments for individual areas, particularly in light of the latest science in this area. It (the BMO) implies that levels of bushfire risks are of similar intensity across the state despite diverse conditions of the vegetation, weather, and topography. Yet, it does allow for more detailed site assessments if considerable expertise and resources were able to be applied via provisions for "alternative methods," suggesting the potential for lowering of risks where appropriate. However, AS 3959 is required as the main requirement for development under the BMO, which in itself does not comply in all cases with the latest science.

The post-2009 development of AS 3959 illustrates the influence of power distribution on decision-making, and, more importantly, the consequences of its unevenness. In this case, power distribution is mainly with the representatives of the timber industry (Professional 1, 2015; Professional 2, 2015; Professional 4, 2015), which has led to the self-removal of the core scientific agency from the process of development of bushfire protection solutions. Taking a step further, this example supports the idea that when two or more institutions are integrated, there is a need to address power relations (Jepperson, 1991; Meyer & Jepperson, 2000) to ensure that power, including the decision-making, is appropriately distributed. Moreover, according to the bushfire professional interviewed:

> *[AS3959] is a risk based assessment, which is good. The problem with AS3959 is that it's a very complex document and there's a strange standards for AS3959. On that standard you've got industry people, you've got fire service people, you've got fire scientists, you've got planners, you've got all these what we call lobby groups and none of them get on. So the people that represent the timber industry will go down a certain path because they want to make sure AS3959 doesn't affect their business. So what we do is we've got this really complex mismatch document that confuses everyone.*
>
> **Professional 2 (2015)**

This evidence demonstrates the strong importance of power distribution in decision-making and the need to find ways to ensure more democratic approaches toward integrated disaster risk reduction (DRR) in the ways it is actually implemented.

The CSIRO professionals involved, however, found another avenue to ensure that scientific evidence is applied in an acceptable manner from their perspective. New building controls for steel framed construction in bushfire areas were introduced in the NASH standard in May 2015 (NASH, 2015). While it is applied using similar decision and testing approaches in parallel with AS 3959, it is predicated around the use of steel framed construction rather than wood. So while the planning methods used remain the same as for AS 3959, this standard "provides a completely new position in design outcome, robustness redundancy and cost effectiveness" (Professional 1, 2015). This example also demonstrates that alternative routes should be sought for applying the latest evidence to ensure a diverse set of views and potential outcomes.

BETWEEN RISK AND POLITICAL REALITY

The allocation of CFA to the role of main consultant agency resulted in a disparity not only between perceived and real fire risks in the state but also in differences between planning and fire engineering perspectives. CFA professionals and volunteers have highly developed professional and practical knowledge of bushfire risks. While their direct involvement in the decision-making process of assessing permits under WMO appears to be logical, their knowledge varied across the state, often not reflecting accurately the overall strategies, and may have hindered balanced risk assessments. Accordingly, permits or refusals issued might not follow overall planning goals and targets that balance out risks.

Another significant drawback of the observed disparity between risks assumed under new policies and community realities was the effective sterilization of large areas of land and the inability of many residents to develop on their land. This has been considered by many as a breach of their human rights (UNISDR, 2015) leading to a significant backlash from the community. This ultimately led to some relaxation of BMO regulations under amendments presented in 2014 and discussed further in this chapter.

Importantly, the need for a permit is exempted when schedules to an overlay such as the BMO specifically state that it is not required. This includes alterations or extensions to existing dwellings less than 50% of floor area; buildings used for accommodation (excluding dwelling and dependant's person unit), which is less than 25% of the existing building area; and buildings with floor areas less than 100 m^2 along with the dwelling used for accommodation (Victoria Planning Provisions, 2011). This exemption, intended as a way of allowing preexisting structures to be maintained without excessive burdens on owners, also appears to be used as a "loophole" by many residents in areas with higher risk levels, meaning that ongoing development is occurring without being subject to bushfire controls. The outcome is that in many cases structures are becoming riskier over time, rather than improving.

THE GROWTH OF GRASSROOTS AND A BACKLASH

While the community backlash in this case is mainly a result of strict new policies that felt punitive to residents, the lack of community involvement in the policy development processes is another potential reason. This argument is based on the recurring comments made by interview subjects that the community was not provided with an explanation of the new policies or residents were not educated on risks and consequences of noncompliance. Another flaw of the new policy was a perceived lack of public consultation and a lack of rights to object or amend the policy to address risks in more realistic manner.

It is possible that the type of consultation undertaken plays a role in community perceptions. Some consultations were undertaken as a part of the long-term recovery stage, when the VBRC carried out 26 community consultations in 14 locations that were significantly affected by fire events. Overall attendance in the period of March 18 and April 9, 2009, was about 1200 people. These consultations aimed to provide firsthand description of people's experiences during events and gain insights as to how DDR practices could be improved. Aside from consultations, public submissions were open to the VBRC for any individual or organization with relevant information. These submissions were extended not only to the country level but also to the international level (Victorian Bushfire Royal Commission, 2009). These consultations, however, were not continued, and the community remained "in the dark" regarding risk levels and development of the detail of statutory policies. As discussed in Chapter 14 by Kate Cotter, community reaction to the policies "pushed" the minister to implement some amendments to the policies after a little more than 3 years of existing in its original form.

After increasing levels of resident opposition and activism in rural electorates, particularly in marginal voting areas, and with a looming election, on July 31, 2014, the Minister for Planning, Matthew Guy, approved further changes to the Victorian bushfire planning laws. The changes were referred to as "practical reforms for building in bushfire designated areas" (State of Victoria, 2014) and sought to ease the process of preparing planning applications affected by the BMO for landowners and to facilitate the process of assessing them for planners in local councils. The changes made to the BMO in 2014 removed some of the more stringent aspects of controls and provided more freedom and performance-based approaches to bushfire risk assessment, its prevention, and mitigation.

The explanatory report of Amendment VC109 outlines that the provisions were meant to:

> *allow landowners to build a new single dwelling on infill lots within an existing subdivision zoned: Neighbourhood Residential Zone, General Residential Zone, Residential Growth Zone, Urban Growth Zone, Low Density Residential Zone, Township Zone and Rural Living Zone; and replace or extend an existing dwelling in an extreme bushfire risk area. In these circumstances, applications will be required to meet an improved bushfire construction standard, but will not be assessed against the broader landscape hazard*
>
> **Government of Victoria (2014, p. 2)**

This and other reasons for change, however, were not considered "truthful" by an independent bushfire engineer, who stated:

> *So BMO 2014 was a political, directly political response to a lobby group's opposition to BMO 2011. And the problem with that is that BMO 2014 is a less rigorous, less logical document. So for example you get, you get a discount for having a private bushfire shelter. So you say okay we're going to look at the risk of design the house to withstand that risk. If you put a bushfire shelter in you can actually line that back one. It's completely illogical.*
>
> **Professional 4 (2015)**

This statement suggests that there might be underlying reasons for some aspects of the detail of the amendments and changes to the BMO in 2014. The rationale for such assumptions can also be found in political and personal influences in planning practices and outcomes, which are set out in a range of wider scholarly literature (Flyvbjerg, 1998, 2001; Hillier, 2002; Stretton, 1989). Ideally, on identifying the reasons for changes in the BMO, implications, both real and possible, should be studied and analyzed to understand positive and negative sides of the process and suggested outcomes. However, due to the limited timeframe of the policy and need in additional development time for it to become evident, it is proposed to create models of possible scenarios of the development under BMO.

The significant changes made to all Victorian planning schemes referred to above as VC109 were introduced using a clause of the Planning and Environment Act (1987) known as Section 20(4). Thus a clause allows the minister to make changes without consultation or opportunities for submissions or appeal, if he/she considers the matter to be of "state significance." These changes were considered to fall into that category due to the impacts that were argued to be affecting the integrity of the planning system in terms of residents' and experts' faith in it.

Table 10.1 demonstrates changes across the three overlays relevant to the study. As it can be noted, there has been a rapid change between 1997 and 2011, with stricter regulations in the latter edition, while 2014 regulations are more relaxed compared to 2011, yet more detailed than 1997.

As can be seen, the changes made in VC109 could be described as minor when compared to the relatively radical changes that were made to facilitate the introduction and widespread application of the BMO after the 2009 fires. However, certain

Table 10.1 Comparison of WMO, BMO 2011, and BMO 2014 Versions

Policy	WMO	BMO 2011	BMO 2014
Defendable space	Not specified	Outer and inner	Inner
Planning integration with other practices	Some	Some	Some
Landscape type as a precondition of level bushfire defense necessary	Not specified	Yes	Yes
Vegetation control	Yes	Yes	Yes
Slope control	Not specified	Yes	Not specified
Equity of application processes	Not specified	Yes	Residential areas have simplified processes
Water supply	Yes	Yes	Yes
House specifications/ design guidelines	Some	Some	Strong policy base
Siting of a dwelling	Specified	Specified	Specified with exemptions
Life span of policy and reasons for change	1997–2011	2011–14	2014-Till date
Consistency of application	No	Yes	Yes
Building to flame zone	Not specified	No	Yes
Role of professionals	Not specified	Veto right and refusal of application	No
Total ban of development	No	Yes	No
Right to appeal	Not specified	Limited	Increased
Alternatives (e.g., bunkers)	Not specified	No	Yes

Developed by Maria Kornakova.

key features are noteworthy insofar as they illustrate political and institutional aspects of this planning provision. First, the new controls were introduced rapidly and with consultation only between limited numbers of stakeholders within key agencies, with the exception of one prominent outspoken activist. This "haste" could be attributed to the government's desire to deal with the highly unpopular matter before impending state elections. Furthermore, the amendment included a number of inconsistencies and minor drafting errors that suggest the normal checking processes were not followed.

In essence, the changes made by VC109 modified key variables in the bushfire tests applied to new developments so that it could no longer be said that the BMO had "sterilized" land. Where previously some townships had many sites on which it was highly difficult or prohibitively expensive to build, VC109 facilitated development on previously subdivided land in high risk areas. This reduced considerably the ire that many rural township residents felt about planning controls that they considered unneeded or inappropriate. This meant that politically, the government had considerably reduced the potential for election campaigns to be run on this basis in country areas.

CONCLUSION

This overview of wildfires in the context of the state of Victoria demonstrates the significance of this natural hazard for the area, outlining some reasons for decision-making oriented to urban planning and building code changes. As stated in the introduction, this chapter shows that postdisaster recovery processes do offer significant opportunities for reforms in certain contexts. In this case the planning and building regulations that were triggered by disastrous events of the 2009 Victorian wildfire season were investigated. Widespread changes were brought about but were primarily the application of existing approaches and understandings of science that were previously underapplied. The 2009 event provided an opportunity to significantly improve the use of these factors, using them in conjunction with a number of other reforms oriented to integration and partial retasking of the activities of a number of agencies. However, these changes were not unproblematic and highlight a number of political, community, and agency culture issues.

The chronological overview of policies and regulations before and after the 2009 Victorian wildfire season provides an overview of changes that were implemented. The development of AS 3959 was the first attempt to formalize and standardize the quality of building structure in the bushfire risk areas. At the time, this standard was developed in the collaboration with CSIRO. This lead Australian agency, however, left the committee in the post-2009 review of the standard as they believed that the decision made was rushed to meet political imperatives and did not comply with the most well-developed science at the time. The development of the BPA introduced the concept of different wildfire risk levels in the state; however, these regulations were still targeted only toward building control.

The implementation of WMO in 1997 was the first exercise of planning controls dealing with wildfire risk areas in the state. While it might seem that community tailored application of this mapping exercise would have been a positive exercise, the reality of the costs associated with it resulted in its inconsistent application and lack of relevant regulations in place when disaster hit the state in 2009. The response of the VBRC to the events resulted in more rigorous and far more widespread application of a simplified version of this approach via bushfire risk area and implementation of statewide planning policies.

Subsequent development and application of the BMO was undertaken as an expert-driven approach that subsequently led to a significant community backlash, in conjunction with CFA officers being required to take on assessments that were often beyond their abilities to judge acceptable risks in a wider way. The subsequent grass-roots backlash brought about ministerial intervention as a form of "adjustment" that found a middle ground to ensure communities could continue to grow without excessive impediment.

Overall, the Victorian case and the changes that were implemented is not only an example of disaster as a trigger of change but also can be considered a relatively successful example of how the political processes can be used as adjustment mechanisms for planning.

REFERENCES

Australian Emergency Management Institute. (2006). *Hazards, disasters and your community: A booklet for students and community*. Retrieved from http://www.preventionweb.net/files/4005_Hazards7thed.pdf: Emergency Management Australia Retrieved from http://www.em.gov.au/Documents/Hazards/Disasters_and_Your/Community.pdf.

Blanchi, R., Leonard, J., Haynes, K., Opie, K., James, M., & de Oliveira, F. D. (2014). Environmental circumstances surrounding bushfire fatalities in Australia 1901–2011. *Environmental Science & Policy*, *37*(0), 192–203. Retrieved from http://dx.doi.org/10.1016/j.envsci.2013.09.013.

Bradstock, R. A., Williams, R. J., & Gill, A. M. (2012). *Flammable Australia: Fire regimes, biodiversity and ecosystems in a changing world*. Retrieved from http://UNIMELB.eblib.com.au/patron/FullRecord.aspx?p=865051.

CFA. (2012). *History of CFA*. Retrieved from http://www.cfa.vic.gov.au/about/history/.

Country Fire Authority. (2012). *About Black Saturday*. Retrieved from http://www.cfa.vic.gov.au/about/black-saturday/.

Farnsworth, S. (2012). *Black Saturday arsonist jailed for almost 18 years*. ABC News. Retrieved from http://www.abc.net.au/news/2012-04-27/black-saturday-arsonist-sentenced-to-28holdholdhold29/3976564.

Flyvbjerg, B. (1998). *Rationality and power: Democracy in practice*. Chicago, IL, USA: University of Chicago Press.

Flyvbjerg, B. (2001). *Making social science matter: Why social inquiry fails and how it can succeed again/Bent Flyvbjerg* (S. Sampson, Trans.). Cambridge: Cambridge University Press.

Gangemi, M., Phillips, S., Stewart, M., Martin, J., & Marton, R. (2003). *A report on the socio-economic impact of bushfires on rural communities and local government in Gippsland and North East Victoria* (Retrieved from Melbourne, Australia).

Gibson, H., Carew, M., Dwyer, M., Jerome, P., & Mitchell, K. (1997). 12.4 special building overlay. Planning and Environment Act 1987. Amendment VC 109 explanatory report. (2014). In: Advisory Committee on the Victoria Planning Provisions (Ed.), *Report of the advisory committee on the Victoria planning provisions (VPPS)*.

Government of Victoria. (2014). *Planning and Environment Act 1987. Amendment VC 109 Explanatory report*. Melbourne: State Government Printer.

Hennessy, K., Fitzharris, B., Bates, B. C., Harvey, N., Howden, S. M., Hughes, L., … & Warrick, R. (2007). *Australia and New Zealand* (Cambridge, UK).

Hillier, J. (2002). *Shadows of power: An allegory of prudence in land-use planning/Jean Hillier*. New York; London: Routledge.

Holland, M., March, A., Yu, J., & Jenkins, A. (2012). Land use planning and bushfire risk: CFA referrals and the February 2009 Victorian fire area. *Urban Policy and Research, 31*(1), 41–54. http://dx.doi.org/10.1080/08111146.2012.736255.

Jepperson, R. L. (1991). Institutions, institutional effects, and institutionalism. In W. W. Powell, & P. J. DiMaggio (Eds.), *The new institutionalism in organizational analysis* (pp. 143–163). Chicago, IL: University of Chicago Press.

Meyer, J. W., & Jepperson, R. L. (2000). The "actors" of modern society: the cultural construction of social agency. *Sociological Theory, 18*(1), 100–120. http://dx.doi.org/10.2307/223284.

NASH. (2015). *National standard of steel-framed housing*. Retrieved from http://www.nash.asn.au/nash/.

Professional 1. (2015). *Interview by Maria Kornakova*. Semi-structured interview, Melbourne.

Professional 2. (2015). *Interview by Maria Kornakova*. Semi-structured interview, Melbourne.

Professional 4. (2015). *Interview by Maria Kornakova*. Semi-structured interview, Melbourne.

Professional 5. (2015). *Interview by Maria Kornakova*. Semi-structured interview, Melbourne.

Professional 6. (2015). *Interview by Maria Kornakova*. Semi-structured interview, Melbourne.

Packham, D. (1992). Bushfires in Australia. *Australian Planner, 30*(1), 8–12. http://dx.doi.org/10.1080/07293682.1992.9657542.

Russell, M. (July 31, 2013). *Black Saturday arsonist sued for compensation*. The Age. Retrieved from http://www.theage.com.au/victoria/black-saturday-arsonist-sued-for-compensation-20130731-2qyg6.html.

State of Victoria. (2014). Practical reforms for building in bushfire designated areas. In Department of Transport Planning and Local Infrastructure (Ed.), *Information sheet* (p. 4). Melbourne: Australia Victorian Government, Department of Transport, Planning and Local Infrastructure.

Stretton, H. (1989). *Ideas for Australian cities/Hugh Stretton* (3rd ed.). Sydney, NSW: Transit Australia.

Victoria Planning Provisions. (1997). *Clause 44.06 Wildfire Management Overlay*. State Government of Victoria.

UNISDR. (2015). *Sendai framework for disaster risk reduction 2015–2030*. Paper presented at the 3rd United Nations World Conference on DRR Sendai, Japan. http://www.unisdr.org/we/coordinate/hfa-post2015.

Victoria Planning Provisions. (2011). *Clause 44.06 bushfire management overlay VC83*. State Government of Victoria.

Victoria Planning Provisions. (2013). *Clause 52.47 bushfire protection: Planning requirements*. State Government of Victoria.

Victorian Bushfire Royal Commission. (2009). *Interim report*. Retrieved from Victoria http://www.royalcommission.vic.gov.au/getdoc/f91ff31a-f477-4543-85ca-ead47f27c58f/Print-Friendly-Version.

Victorian Bushfires Royal Commission. (2010a). *Final report. Summary*. Victoria: Government Printer for the State of Victoria.

Victorian Bushfires Royal Commission. (2010b). The fires and the fire-related deaths. In B. Teague, R. Mcleod, & S. Pascoe (Eds.), *The 2009 Victorian Bushfires Royal Commission final report*. Victoria: Government Printer for the State of Victoria.

Victorian Bushfires Royal Commission. (2010c). Planning and building. In B. Teague, R. Mcleod, & S. Pascoe (Eds.). B. Teague, R. Mcleod, & S. Pascoe (Eds.), *Fire preparation, response and recovery:* (*Vol.* 2). Victoria: Government Printer for the State of Victoria.

CHAPTER

11

Toward Sustainable Disaster Recovery by Seeing It as "More Than a Roof Overhead"*

Esther Charlesworth[1], Iftekhar Ahmed[2]

[1]RMIT University, Melbourne, VIC, Australia; [2]University of Newcastle, Callaghan, NSW, Australia

MORE THAN A ROOF OVERHEAD

Disasters present extremely challenging circumstances to humanitarian agencies and professionals. The tasks of postdisaster reconstruction and recovery are demanding and include a wide range of activities as clearing and sorting the vast amounts of debris and rubble for possible reuse; reestablishing water, power, and waste services; ensuring temporarily housing the severely displaced; and planning their long-term homes while dealing with the intense grief and psychological traumas of the disaster victims. These challenges continue to become more complex with the ongoing global changes of urbanization, environmental degradation, and climate change. What constitutes *the* "disaster," however, should not be seen as limited to just the physical and financial damage left by the increasing number of floods, tsunamis, earthquakes, and typhoons, such as that we have witnessed globally since the 2004 Indian Ocean tsunami, one of the most severe disasters in recent history, which again brought a global focus on these events. Underpinning these impacts and, indeed, exacerbating them are sources of social and physical vulnerabilities, such as systemic poverty, political neglect, inappropriate zoning, altered land uses and landscapes, and a lack of long-term investment in adequate health and educational infrastructure. Such factors are critical in framing both the problems of, and successful approaches to, rebuilding vulnerable communities.

The impacts of an extreme (un)natural disaster are most commonly measured in terms of lives lost, houses damaged and destroyed, people displaced and homeless, and the consequent financial cost (UNISDR, 2015). With shelter being an immediate need, housing—from emergency to transitional and permanent shelters and homes—is a primary and early task for humanitarian and recovery workers. Unfortunately,

*The case studies in this chapter were drawn from a book by the authors (Charlesworth & Ahmed, 2015). The research for the book was originally funded by a Future Fellowship awarded by the Australian Research Council to Charlesworth in 2011.

Urban Planning for Disaster Recovery. http://dx.doi.org/10.1016/B978-0-12-804276-2.00011-6

and very often, the housing that results is often inappropriate in design, unlivable in terms of local climatic conditions, built with inappropriate and often imported materials, fails to provide employment for displaced people during construction, and neglects the services and industries needed for long-term recovery (Haiti Grassroots Watch, 2014; Sadiqi, Coffey & Trigunarsyah, 2012; Schuller, 2016). This is despite copious advice on policies and strategies for building resilience during recovery and guidelines for sustainable community development and housing construction issued by international agencies and academic researchers (Davis & Alexander, 2016; IFRC, n.d.; Lyons, Schildermann & Boano, 2010).

The term "more than a roof overhead" is taken from Newman (2002) who argued that the provision of housing for marginalized and displaced people, indigenous ones in his case, should seek to provide both shelter and nonshelter outcomes, with the latter including opportunities for family stability, improved health and education, and employment. Acquilino (2011) uses the term "beyond shelter" to connote the same meaning, whereas Davis and Alexander (2016) uses the term "holistic and integrated recovery." This chapter extends Newman's metaphor to argue that postdisaster recovery should aim to provide not only shelter but also the built, social, and economic infrastructure to mitigate preexisting vulnerabilities and support successful long-term recovery—the dual goals of "building back better."

This chapter uses two case studies to illustrate how addressing the intersection of the sources of vulnerability and the impacts of a disaster can frame the processes of recovery. The case study communities, a village in Sri Lanka and a city in the United States, were both severely impacted by (un)natural disasters after long periods of poverty and neglect. The case studies have been selected to demonstrate the complexity of housing reconstruction and the importance of approaching reconstruction challenges through the lenses of "more than a roof overhead." This, perhaps, is what sustainable disaster recovery means.

The case studies suggest that success will be limited if the recovery of vulnerable communities focuses solely on rebuilding houses without adequate attention being also paid to rebuilding the social and physical infrastructure of a community. When reconstruction projects achieve sustainable outcomes—that is, when communities and not just buildings are rebuilt, when infrastructure needs are considered at the same time as housing needs, and when people make genuine input into the design of their homes and communities—these are the types of projects that can offer valuable lessons for action following future disasters.

The case studies examined in this chapter are:

1. **FoG Villages, Sri Lanka**: After the 2004 Indian Ocean tsunami, the Foundation of Goodness (FoG) undertook a housing resettlement and civic infrastructure program in Seenigama on the southern coast as part of a long-term community development initiative for communities severely affected by the tsunami.
2. **Musicians' Village, USA**: Hurricane Katrina in 2005 spurred a participative reconstruction program by Habitat for Humanity in New Orleans, building a Musicians' Village to resettle the musicians of this city with a strong musical heritage.

While implemented in different geographic contexts and by different agencies, both projects were atypical of the many projects initiated in southern and eastern Sri Lanka and New Orleans and other Gulf communities. In terms of "more than a roof overhead" success in both these projects, it resulted from linking the physical products of reconstruction to a wider set of community and institutional processes. The next two sections of this chapter provide the case studies. These are followed by a conclusion that synthesizes the challenges encountered in this approach.

CASE STUDY 1: RECONSTRUCTION OF A TSUNAMI-AFFECTED COMMUNITY, SEENIGAMA, SRI LANKA

Context/Crisis	Indian Ocean tsunami, 2004
Location	Seenigama, Galle District, Sri Lanka
Type of project	Resettlement and housing reconstruction of tsunami-affected communities
No. of houses	153 houses of varying designs
Cost per m^2	Sri Lankan Rupees 8000 (about US$80)
Date completed	December 2007
Implementing agency	Foundation of Goodness (FoG)
Donors	*Aviva Village*: AVIVA-UK and WNS Customer Solutions—Sri Lanka *Perth Village*: Perth City Council, Australia *KPMG-LOLC Village*: KPMG-UK and LOLC-Sri Lanka *Victoria Gardens*: State Government of Victoria, Australia

The village of Seenigama, near the towns of Galle and Hikkaduwa, is located on the southern coast of Sri Lanka. Its low-lying landscape coupled with the mining of limestone from the protective near-shore coral reefs led the tsunami to have a particularly catastrophic impact on the district. The recovery process in Seenigama was unique in Sri Lanka in that the implementing agency, FoG, was already situated within the affected community. FoG was founded in 1999 by Kushil Gunasekera, the son of a local landowner. Before the Indian Ocean tsunami in 2004, FoG provided community services, such as English and computer training, sports opportunities, a maternity clinic, and water and electricity supplies. In addition to family funds, Gunasekera was able to catalyze several international firms and agencies—from his business and sporting contacts—to support the various community development initiatives that FoG initiated.

After the devastation of the Indian Ocean tsunami, FoG marshalled previous partnerships, as well as new ones, to rebuild the Seenigama community. Four resettlement "villages," consisting of 153 houses and supporting community infrastructure and services, were constructed in the area surrounding FoG headquarters in Seenigama.

In all, FoG built 625 new houses and repaired 401 others in Seenigama and nearby villages. This case study deals with the villages around the FoG headquarters.

In most postdisaster housing reconstruction projects, not only after the tsunami in Sri Lanka but also, indeed, in most parts of the world, it is common for implementing agencies to be external to the country or based away from the project site. They tend to stay for a limited time to implement their projects and then move on to new humanitarian emergencies elsewhere (Alexander, 2013). On the other hand, FoG was a local nongovernmental organization under Sri Lankan leadership and deeply embedded within the local culture and economy. This allowed FoG to readily appreciate local needs, to consult widely and effectively about local goals for recovery, and to build on the trust and social capital it had been instrumental in establishing. It also enabled FoG to sustain its work beyond the postdisaster relief and transitional stages of reconstruction.

Additionally, FoG's work encompassed a wide range of initiatives beyond the construction of new housing. Following an integrated postdisaster community development approach, FoG was able to address predisaster vulnerabilities to support long-term development and resilience. For example, when an international telecommunication company offered to fund a training center in Seenigama, FoG negotiated for scuba and underwater construction (e.g., of piers and ports) to be the focus of training, thus providing alternative employment for the coral divers who had been driven by lack of other opportunities and skills to damage the local fringing reefs. As a result, the further weakening of the reef as a safety barrier was prevented, stabilizing physical infrastructure for resilience. A women's training center, specializing in office and computer skills, home and childcare, retail sales skills, and textile crafts, was also established as a kindergarten and a national center for sporting excellence (which drew further financial support from overseas sporting associations and clubs).

The four "villages," of varying sizes and house designs that were constructed, were conceived within this community development approach. For the largest village, Victoria Gardens, Global Modular Housing Pty Ltd, a Melbourne-based housing supplier, was contracted by the Victorian Government to design and manage the project. However, the site plan was developed by DPM Consultants, Sri Lanka, led by their Principal Architect Jayantha Bandara. Various Sri Lankan–based architects, engineers, and subcontractors were appointed throughout the Victoria Gardens and other village projects.

The four posttsunami housing "villages" built by FoG in Seenigama include:

- **AVIVA Village**: A group of 10 single-storey detached houses supported by a community center, computer and English training center, library, a maternity clinic, and a playground. This village was funded by AVIVA-UK and WNS Customer Solutions-Sri Lanka.
- **Perth Village**: This was a cluster of nine single-storey detached houses next to the AVIVA Village, which benefitted from the community facilities of AVIVA Village. This village was funded by the Perth City Council, Western Australia.

- **KPMG/LOLC Village**: This was a group of 50 single-storey detached houses together with a community center, a water supply tower, a library, and a playground. In all, 25 houses were funded by KPMG-UK and 25 houses by LOLC-Sri Lanka.
- **Victoria Gardens**: This was a planned settlement of 84 two-storey duplexes laid out around a central community area that included a community center, a water purification plant, a sewage treatment plant, and a playground. The project was funded by the State Government of Victoria, Australia. This was FoG's largest resettlement project and was built on newly acquired land to rehouse tsunami-affected households that had lost their coastal property because a government declared posttsunami "buffer zone" policy prevented rebuilding close to the coast.

A variety of house types were built in the FoG projects. The earliest houses were single-storey models, the evaluation of which provided lessons for the building of Victoria Gardens. For example, one issue that was particularly significant came to be a preference for two-storey houses. Interviews with some of the residents of the first three villages found that they would prefer a two-storey house due to fear of future tsunamis. Thus, several house designs were offered to potential residents of Victoria Gardens and after wide discussions the designs selected comprised two-storey, two-bedroom houses in a duplex arrangement (see Fig. 11.1). This design option gave the residents greater safety (structurally strong concrete slab and wall construction, with the upper floor level above the height of the recent tsunami) should they experience another tsunami.

FIGURE 11.1

A view from Victoria Gardens showing the housing laid out around a central community area and playground.

SUCCESS FACTORS

The success of the projects in Seenigama was driven largely by the strong role played by FoG and its long-term engagement with the community. These resulted in five foundations for physical and social resilience and included the following:

- *Diversity in housing layout*: Each of the four "villages" had a unique character, with a clustering of similar houses. Although the house plans were similar within each village, site planning allowed for variety. In Victoria Gardens, for example, the duplexes had different orientations, different external color schemes, and different roof shapes.
- *Community infrastructure*: The houses were built as part of a system of community infrastructure that supported the important physical and social needs of the community. This included paved roads within the sites; power and street lighting; reticulated treated water supply from deep wells on-site; and a reticulated sewerage system, including an on-site tertiary treatment plant in Victoria Gardens.
- *Maintenance*: Unlike most agencies that had implemented posttsunami housing reconstruction projects and then left the beneficiary community behind, FoG continued to support the maintenance and upkeep of its housing and infrastructure projects. Communal areas such as playgrounds and open spaces were regularly maintained by mowing the grass and trimming bushes; cleaning the drains; and repairing roads as required. This sustained and contributed to consolidating the resilience of the community and the housing system.
- *Embedded within the community*: After the many international agencies, which came to assist Sri Lanka, had completed their housing projects, they concluded their operations in the area. Very few maintained any link with beneficiary communities. FoG, on the other hand, was embedded within the community and, after implementing the posttsunami housing and infrastructure projects, continued to support the community with a range of neighborhood services.
- *Integrated community development*: The final and, perhaps, key success factor was the integrated community development approach followed by FoG. There was clear understanding that building houses alone was not sufficient; housing had to be backed and sustained by a range of elements, both physical (e.g., roads, water, electricity, sanitation) and social (e.g., education, livelihoods, sports). The FoG housing projects had been implemented with the infrastructure necessary for community functioning and services and were sustained by a wide range of community development activities. Located around the FoG building, such activities included training, employment and cultural services, such as computer and English training, and other forms of vocational and business skills training—preschool, medical center, library, scuba diving training center, and sales outlet for products made by women with FoG's support. Sport was seen as an integral part of children and youth development, and a variety of opportunities were provided within the Seenigama community, including a swimming pool, cricket ground, gymnasium, and a training center of excellence for youth (Fig. 11.2).

FIGURE 11.2

Beneficiaries of FoG who with a house regained their livelihoods—traditional dancing lessons and a grocery shop run from the house.

CASE STUDY 2: MUSICIANS' VILLAGE, NEW ORLEANS, USA

Context/Crisis	Hurricane Katrina, 2005
Location	Upper Ninth Ward, New Orleans, Louisiana, USA
Type of postdisaster project	Integrated housing, recreation, and music center
No. of houses	72
Cost per m^2	US$100,000 per house Total cost including duplexes for the elderly, a children's park, and the Ellis Marsalis Music center is US$20,726,500
Date completed	2008
Implementing agency	New Orleans Area Habitat for Humanity (NOAHH)
Donors	Funds contributed by NOAHH and the First Baptist Church and also raised from various other sources

When Hurricane Katrina struck the US Gulf Coast on August 29, 2005, the city of New Orleans, Louisiana, felt the maximum brunt of the fierce winds and flooding and the impacts extended for hundreds of kilometers along the coast to the east and west, as well as inland to the north. Hurricane Katrina brought about the deaths directly or indirectly of more than 1800 people, almost 80% of whom were in New Orleans. Katrina also injured more than 5000 others and caused damage worth over US$100million. More than one million people were displaced. Over one million housing units were damaged, more than half of which were in Louisiana. Mississippi was the next most affected state where more than 220,000 houses were damaged (see Blake, 2011; Liu, Anglin, Mizelle, & Plyer, 2011).

The extensive destruction of dwellings made rehousing a priority for those who had suffered property loss. Initially, temporary accommodation was provided in caravans ("FEMA trailers") by the Federal Emergency Management Authority. These were, at best, a short-term solution and the development of permanent housing remained a primary need. One of the first design solutions was "Katrina Cottage." This was around the same size as a FEMA trailer (28.6 m^2) and was made of lightweight materials that could be easily assembled on-site. With plans and provisions for future extension, this became a widely adopted housing option (Alter, 2008). The Katrina Cottage design also influenced the housing built in the following case study of the Musicians' Village.

However, housing reconstruction efforts post-Katrina were quite diverse with many significant initiatives from the government, business, and philanthropic sectors. One initiative was the Make It Right project. Initiated and funded in 2007 by the actor Brad Pitt, Make It Right engaged 21 (mostly) renowned architects to design houses for residents from one of the most ravaged areas of New Orleans, the Lower Ninth Ward (see Feireiss, 2009). Incorporating a variety of innovative designs, the houses were well built, responsive to flood risk, and energy efficient. By 2013, 90 out of the planned 150 houses had been built. However, the architects chose different design forms and construction systems, thus raising the price of many of the houses beyond the reach of most of those who that lived in the area pre-Katrina (DePillis, 2013). This made bringing people back to the Lower Ninth Ward a significant challenge, along with insurance and proof of tenure issues for many. After Katrina many of the residents left, and with the stigma of being a poor neighborhood lacking in infrastructure and amenities, the revitalization of the area was difficult.

In fact, poverty was widespread in New Orleans before the hurricane, with African–American communities among the poorest. Many of them, including many of the musicians in this city of a longstanding music tradition, were concentrated in the lowest, most flood-prone areas such as the Lower Ninth Ward (Logan, n.d.). Not only these communities were the most exposed and vulnerable to the hurricane, but low levels of car ownership among them, poor public transport, and the lack of adequate evacuation plans also made them vulnerable among the most severely impacted. The Musicians' Village project was among a number that targeted the reconstruction needs of such communities. Instead of only building houses, the project aimed to build a community, indicated by its name, "Musicians' Village." This was linked to its other aim of preserving the local music heritage, thus going beyond only housing reconstruction.

The idea of building a Musicians' Village in the Upper Ninth Ward came from New Orleans musicians, Harry Connick Jr. and Branford Marsalis. The New Orleans Public Schools Board played an important part by selling the land at nominal cost while substantial funding came from philanthropic donors, musicians who played concerts for the project, and two local community organizations. A year before Katrina, the First Baptist Church of New Orleans began plans to build 40 houses in the area under the name of the Baptist Crossroads Project (BCP). After Katrina, BCP merged with the NOAHH to implement the Musicians' Village project.

The Musicians' Village encompasses five city blocks and consists of 72 detached houses, 5 duplexes for elderly residents, the Ellis Marsalis Center for Music, and a children's park (Fig. 11.3). The detached houses, whose style often

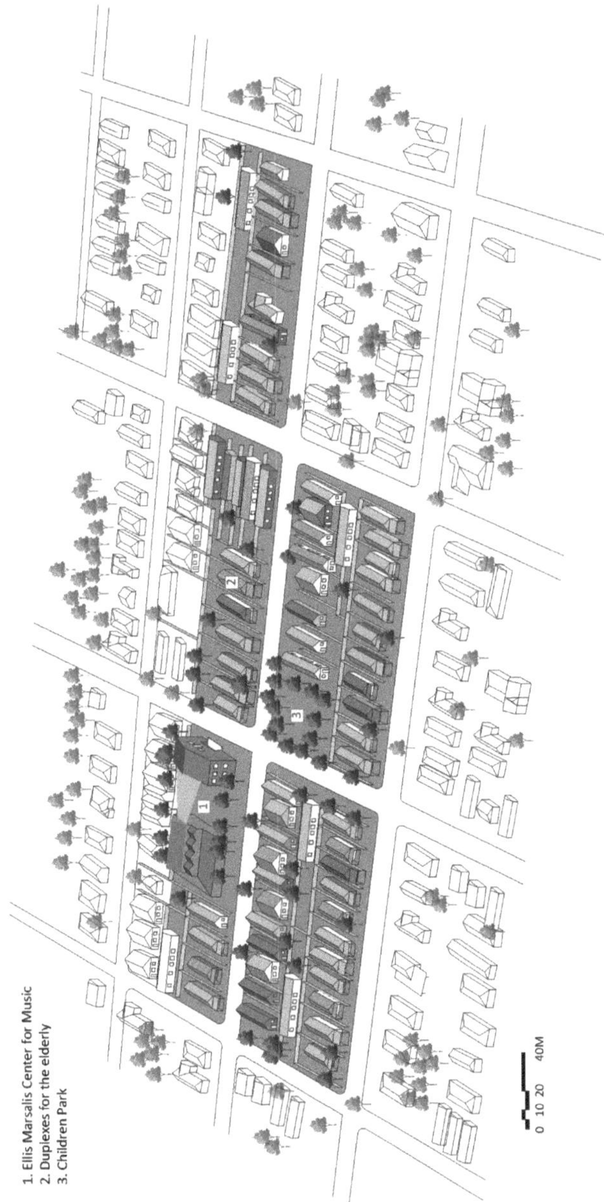

FIGURE 11.3

Sketch showing layout of the Musician's Village.

Courtesy of Kate Ryle

resembled the traditional "shotgun house" design of southern United States, were for sale on 20-year, no-interest mortgages to musicians and their families. Purchasers also had to make a 350-h community service or "sweat equity" contribution. Approximately 100 sq.m. in size, each house offered residents a choice of facades and the opportunity to paint the exteriors and interiors with colors of their choice. There was also an opportunity to make small adjustments to the room layout and position of doors and windows, as well as the colors of carpets, countertops, trim colors, and tiles.

Five duplexes were built for renting to elderly or retired residents. These were grouped facing a side street, with each duplex having two residential units with separate entrances. The duplexes followed codes and standards for elderly and disabled people and included wide doors for wheelchair access, grab rails, and accessible positioning of electrical outlets. These homes were built by more than 120,000 Habitat for Humanity volunteers coming from many different places to support the project.

A children's park was also designed and constructed as a central community facility. In addition to children's recreation (child-friendly equipment and soft landscaping to prevent injury), the park also offers a meeting and relaxation space for adults.

The Ellis Marsalis Music Center was built for the education and performance of local musicians. It is centrally located in the Musicians' Village and provides over 1500 sq.m. of classrooms, recording studios, and a 170-seat performance theater. While it caters mainly to residents of the Musicians' Village, it is open to other musicians and, thus, serves as an important hub of New Orleans's musical heritage (Fig. 11.5).

SUCCESS FACTORS

The success of the Musicians' Village project was due to a combination of factors that contributed to physical and social resilience and included the following:

- *Cultural cohesion*: The design of houses was derived from the style of vernacular New Orleans housing. This made them attractive to potential residents. The many choices that allowed homeowners to personalize the houses helped to develop a sense of ownership, while the goal of having 80% of residents involved in the music industry gave a sense of purpose and direction for community life. The wide range of age groups—from children to the elderly—that were catered for is added to community cohesion. The involvement of residents and volunteers in the construction of the project contributed to the enrichment of social capital within the Village and with people from other parts of the city.
- *Integration with community facilities*: Although the focus was on housing, community facilities such as the music center, elderly residences, children's park, and a thrift shop allowed the project to demonstrate the value of being "more than a roof overhead" (Fig. 11.4).

FIGURE 11.4

A house of a rhythm-and-blues piano player in the Musician's Village.

FIGURE 11.5

The Ellis Marsalis Center for Music, a focal point of the community.

- *Disaster risk reduction*: The project area was flooded by Katrina. However, the construction of a new levee nearby helped to create a safe place to build. Houses were also built more than 30 cm above the FEMA base flood level, which involved a raised platform floor and a flow-through concrete block base for all homes. International building standards for wind-resistant structures and windows were also followed. This meant that there was no structural damage to the Musicians' Village from Hurricane Isaac in 2012 although nearly 60,000 houses were damaged across southeastern Louisiana.

- *Support to the local economy*: Even though there was no direct economic support component other than the "soft" loan terms, the large number of volunteers who came to build the project and stayed near the site brought income to local businesses. For example, two restaurants in the vicinity flourished during the construction. In addition to the sale of building materials, it was estimated that nearly 3500 jobs had been created by NOAHH in New Orleans since 2006.

KEY LESSONS

The Seenigama and Musicians' Village projects offer valuable lessons for using the "more than a roof overhead" approach to postdisaster reconstruction. While some of these lessons are context specific, there are also many that have wider relevance to housing reconstruction after disaster.

- **The first lesson concerns the key message of the need to integrate housing with community infrastructure**:
  Postdisaster reconstruction must go beyond building new houses to replace those damaged to an integrated approach where a range of other elements are provided, particularly community infrastructure and facilities such as roads, water, sanitation, electricity, schools, community buildings, and parks. In Sri Lanka and in the United States, because of the vastly differing level of public services and infrastructure, this was not a key undertaking of the implementing agencies; nonetheless in the Sri Lanka FOG project, community infrastructure, including roads, utilities, playgrounds, and community buildings, was a vital element.
- **Secondly, reconstruction should support livelihoods and the local economy**:
  The opportunity to earn a living is essential for those affected by disaster. Throughout discussions with disaster-affected communities during fieldwork for this research, the regeneration of livelihoods was emphasized as being as great a need as housing. In many communities, the house can also be a workplace for home-based livelihoods, especially for women. Most of the implementing agencies supported livelihoods as part of their housing reconstruction initiatives, through mechanisms such as skills training, provision of equipment, the necessary infrastructure for a livelihood, start-up supplies, or through cash-for-work and/or sweat equity schemes. In addition, the reconstruction projects supported local economies through the creation of jobs and marketing opportunities for a range of local building product suppliers and producers.
- **An important third lesson concerns the need for widespread multistakeholder engagement**:
  The complexity of postdisaster reconstruction demands the involvement of a wide range of stakeholders and professionals in projects. Associated with this is the involvement of community organizations that have been in the affected areas for long periods. They understand community history and culture, are

known and trusted, and are skilled in working in facilitative ways. This brings choice, a sense of direction, a sense of ownership, and a sense of ongoing involvement and responsibility. An increasing emphasis on multistakeholder engagement can be expected in the future as disasters become more complex and the global forces of climate change and urbanization continue to create unprecedented challenges.

- **A fourth lesson concerns the issues of replicability and upscaling**: A key question, articulated by a staff member of one of the implementing agencies in this study, was, "Are we creating islands of benefit in a sea of widespread deprivation by such projects?" These case study projects do indicate possible future directions for effectiveness in rebuilding after a disaster, but there is yet much to be done in widely replicating and scaling up their successes.

CONCLUSION

The analysis above of postdisaster housing reconstruction is drawn from extensive empirical field-based research across two countries that examined the performance and outcomes of postdisaster housing reconstruction projects in countries from the Global South and North. The lessons drawn from the two case studies aim to assist agencies working on future postdisaster reconstruction projects. By focusing on housing reconstruction alone, without the complementary rebuilding of civic infrastructure and the strengthening of local livelihoods and capacity, ultimately leads to very limited long-term development outcomes. As the title of this chapter suggests, building "more than a roof overhead" is critical to the long-term rebuilding of resilient communities after disaster.

REFERENCES

Acquilino, M. (2011). *Beyond shelter: Architecture and human dignity*. New York: Metropolis Books.

Alexander, J. (2013). *Chasing chaos: My time in and out of humanitarian aid*. New York: Broadway Books.

Alter, L. (2008). *Katrina cottages rolled out by Lowes nationwide*. Retrieved from http://www.treehugger.com/modular-design/katrina-cottages-rolled-out-by-lowes-nationwide.html.

Blake, E. S. (2011). *The deadliest, costliest, and most intense United States tropical cyclones from 1851 to 2010 (and other frequently requested hurricane facts)*. Miami: National Hurricane Center.

Charlesworth, E., & Ahmed, I. (2015). *Sustainable housing reconstruction: Designing resilient housing after natural disasters*. Abingdon, UK: Routledge.

Davis, I., & Alexander, D. (2016). *Recovery from disaster*. Abingdon, UK: Routledge.

DePillis, L. (March 13, 2013). *If you rebuild it, they might not come: Brad Pitt's beautiful houses are a drag on New Orleans*. New Republic.

Feireiss, K. (Ed.). (2009). *Architecture in times of need: Make it right rebuilding New Orleans' Lower Ninth Ward*. Munich: Prestel Verlag.

Haiti Grassroots Watch. (2014). *Question about the reconstruction's housing projects*. Retrieved from http://haitigrassrootswatch.squarespace.com/haiti-grassroots-watch-engli/2014/1/8/questions-about-the-reconstructions-housing-projects.html.

IFRC. (n.d.). *From crisis to recovery*. Retrieved from http://www.ifrc.org/en/what-we-do/recovery/.

Liu, A., Anglin, R., Mizelle, R., & Plyer, A. (2011). *Resilience and opportunity: Lessons from the U.S. Gulf Coast after Katrina and Rita*. Washington, DC: Brookings Institution Press.

Logan, J.R. (n.d.). *The impact of Katrina: Race and class in storm-damaged neighborhoods*. Retrieved from www.s4.brown.edu/katrina/report.pdf.

Lyons, M., Schilderman, T., & Boano, C. (Eds.). (2010). *Building back better: Delivering people-centred housing reconstruction at scale*. Rugby, UK: Practical Action Publishing.

Newman, P. (2002). *Sustainability and housing: More than a roof overhead* Barnett Oration, 31 October. Retrieved from http://www.ahuri.edu.au/downloads/2002_Events/Oswald_Barnett/Peter_Newman.pdf.

Sadiqi, Z., Coffey, V., & Trigunarsyah, B. (2012). Rebuilding housing after a disaster: Factors for failure. In *Paper to 8th annual international conference of the International Institute for Infrastructure, Renewal and Reconstruction (IIIRR), Kumamoto, Japan*. Retrieved from https://www.researchgate.net/publication/265707613_Rebuilding_Housing_after_a_Disaster_Factors_for_Failure.

Schuller, M. (2016). *Humanitarian aftershocks in Haiti*. New Brunswick: Rutgers University Press.

UNISDR. (2015). *Global assessment report 2015 (GAR 2015)*. Retrieved from http://www.preventionweb.net/english/hyogo/gar/2015/en/home/.

CHAPTER

Resilient Housing Reconstruction in the Developing World

12

Iftekhar Ahmed
University of Newcastle, Callaghan, NSW, Australia

INTRODUCTION

Housing is often the most valuable and important asset for many people, and its principal role is to provide protection from the elements of nature. Disasters throughout the world often impact severely on housing, and it is usually the most visible element that is damaged or destroyed. Rapid onset disasters such as earthquakes and cyclones cause significant devastation to housing, often leading to loss of this valuable asset; slow onset disasters such as floods and bushfires often displace people from their homes and can also cause destruction (CMC, 2015). Displacement or loss of housing makes people vulnerable to possible aftershocks or to the climate—rain, snow, heat, etc.—compounding the effects of the disaster and hence impacting household and community health. Thus, there is the need for resilient housing to safeguard people from disaster risks and impacts.

Developing countries[1] tend to bear the brunt of disaster impacts, with the poor there often being the most severely affected (Hillier & Nightingale, 2013). Asia, the continent with the highest population and where the majority are developing countries, experiences the greatest disaster impacts in terms of number of disasters, economic damage, and numbers of people killed and affected (ADRC, 2014). Taking only one year, 2010, there were 144 recorded disasters in 30 Asian countries (ADRC, 2012). Table 12.1 shows some selected developing countries where recent disasters have devastated large numbers of housing, indicative of the precarious and life-threatening living circumstances of the great majority of the world's people.

Because of the physical nature of housing, it is particularly vulnerable to disasters, often representing the greatest share of loss in the total impact of a disaster on the economy (Lyons, 2009). For example, in the 2004 tsunami and earthquake in Indonesia, one of the biggest disasters in recent times, the housing sector experienced

[1] The appropriateness of the term "developing" country or world has been debated and alternative terms—Global South, Majority World, etc.—have been suggested by different parties. It has been used here because of its widely understood meaning to signify countries with low socioeconomic and human development levels.

Urban Planning for Disaster Recovery. http://dx.doi.org/10.1016/B978-0-12-804276-2.00012-8

Table 12.1 Impact of Disasters on Housing in Selected Countries

Country	Disaster	Year	Housing Destroyed/ Damaged	Sources
Bangladesh	Cyclone Aila	2009	616,000	IFRC (2010a)
Fiji	Cyclone Winston	2015	31,200	UNOCHA (2016)
Haiti	Earthquake	2010	293,383	United Nations (2012)
India, Indonesia, Maldives, Seychelles, Somalia, Sri Lanka, and Thailand	Indian Ocean earthquake and tsunami	2004	414,935	Disaster Watch (2006)
Nepal	Earthquakes	2015	785,075	UNOCHA (2015)
Pakistan	Floods	2010	1,608,185	ADB and World Bank (2010)
Philippines	Typhoon Haiyan/ Yolanda	2013	1,079,452	Pacific Disaster Center (2013)

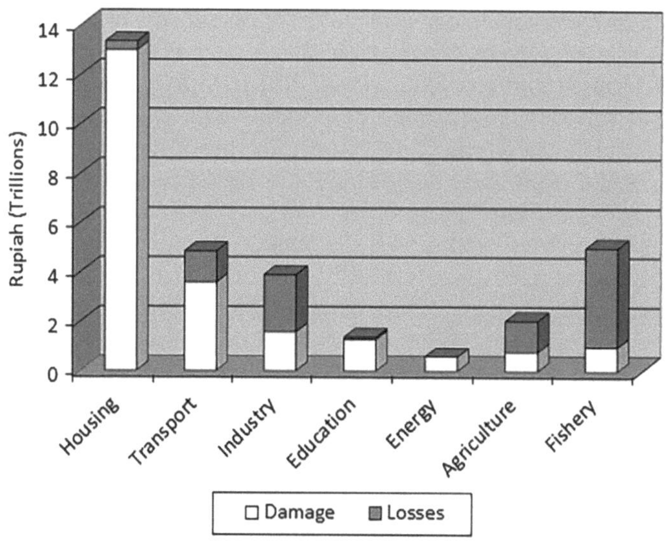

FIGURE 12.1

Impact of the 2004 Indian Ocean tsunami on different sectors in Indonesia; note the proportionately much higher impact on the housing sector.

Adapted from Marti (2005).

maximum economic damage (Marti, 2005) (see Fig. 12.1). In developing countries particularly, the impact of disasters on the built environment is much higher than in developed countries, estimated at more than 20 times in magnitude (Barakat, 2003). Many agencies from prominent bilateral and multilateral bodies to grassroots nongovernmental organizations have begun to take disaster resilience[2] seriously; however, this is yet to converge more strongly with the housing sector in developing countries.

POSTDISASTER RECONSTRUCTION: AN OPPORTUNITY FOR BUILDING RESILIENCE

The increasingly frequent and massive disasters in recent history have necessitated large and extensive reconstruction programs by a wide range of agencies worldwide, particularly after the Indian Ocean Tsunami in 2004. Housing in developing countries is often most visibly and extensively devastated by disasters and thus in many postdisaster recovery programs, the majority of resources and main priority is allocated to shelter and infrastructure reconstruction compared to other sectors (Lang, 2008). Most country-specific guidelines and initiatives for safer buildings have arisen after major disasters such as earthquakes and tsunamis (ERRA, 2006; NHDA, 2005). In places where disasters are frequent and recurrent, such as floods in Bangladesh, or typhoons in Vietnam, resilient housing initiatives are high on the agenda of agencies (Ahmed, 2005, 2016; Ahmed & Charlesworth, 2013; Tro, 2011).

During reconstruction, there is the opportunity to understand and thereby address and overcome the underlying vulnerabilities that had previously prevented resilient housing construction and the risks that threaten durability and sustainability of housing. Based on local knowledge and participation, building housing back to a better standard that is less vulnerable to context-specific hazards can contribute to reduced disaster risks in the long term (Lyons, Schilderman, & Boano, 2010). The involvement of people from the area allows understanding the nature of local risk and accordingly design housing that mitigates the risk. Reconstructed or rehabilitated housing with future risk in mind can prove more resilient. To provide a few basic examples: incorporating earthquake-resistant construction elements such as bracings and struts can reduce future earthquake risk; or building raised shelter in flood-prone areas can protect them from damage.

WIDER LINKAGES

In addition to the above-mentioned technical aspects of resilience, it is important for agencies implementing housing reconstruction projects to take into account a range of social, cultural, political, environmental, and other issues, requiring a sustainable and holistic approach to building disaster resilient housing. As opposed to merely

[2] Resilience is a broader term including a wide range of shocks and stresses (Rockefeller Foundation, 2016), whereas disaster resilience is concerned with the capacity of people and systems to withstand and/or to resist the impacts of disasters and be able to recover easily (UNISDR, 2009). It includes physical "hardening" (Valdes, Amaratunga, & Haigh, 2013) as well as softer aspects such as social capital (Aldrich, 2012).

constructing houses via the provision of funding, agencies need to ensure that design and technical support is provided and the construction is supervised and monitored—and also be aware of constraints that may prevent the implementation and uptake of resilient construction methods. The network of relationships within communities, contributing to their social capital, can often be a key element in resilient postdisaster reconstruction and recovery, as argued by Aldrich (2012) in a variety of contexts.

A study on mainstreaming disaster risk reduction (Wilderspin, Barnham, Gill, & Lockwood, 2008), where the author was a team member, highlighted that during housing reconstruction the main thrust of the intervention should be to build back better so that the repaired or new housing is safer and more resilient, and at the same time local capacity is developed for constructing, replicating, and maintaining such housing. This should be implemented within a framework of local risk assessment, improvement of local building practices and skills, support to local industry and employment, development of improved and more resilient building materials and techniques, and preparedness and contingency planning for subsequent disaster events. Agencies implementing housing reconstruction projects need to be aware of the balance between the provision of housing and ensuring technical support and capacity building. Together with technical assistance, attention has to be paid to ensure a strong degree of skills transfer and development, leading to building the capacity of homeowners, local professionals, and builders, as well as increased market opportunities for local suppliers and construction workers.

LIVELIHOODS: A VITAL LINK

The link between housing and livelihoods is important and contributes to the sustainability of a housing project, especially to achieve disaster resilience. Often evident in postdisaster situations, affected people tend to prioritize housing as their most urgent need together with livelihood regeneration (Delaney & Shrader, 2000; Skinner, 1991). Beside a house being a workplace and having strong implications for health and well-being contributing to economic productivity (HFHA, n.d.a), production of housing after a disaster can create local jobs and regenerate the local economy through production, procurement, and transport of building materials (Cosgrave, 2008; Feinstein International Center, 2011). A study showed that households whose homes were rebuilt after a disaster were able to resume income-generating activities, which allowed economic recovery from the disaster's impacts (HFHA, n.d.b). Importantly, local capacity can be developed for building resilient housing and in this sense allows the disaster resilience initiative to be sustained over the long term.

SOCIAL CAPITAL: THE MISSING ELEMENT

The physical aspects of resilient housing rely strongly on the links that connect people, that is, social capital. Implementation of reconstruction projects requires relationships between institutions and communities, and the involvement of different stakeholders.

Social capital is the glue that joins them to be able to work toward the common goal of building resilience. This is such an important and intrinsic element, but it often gets hidden beneath the discussions on resilience that focus on making stronger buildings and infrastructure, which of course would not be possible without the interrelationships between people.

Social capital has been defined widely in a variety of contexts; for example, Lucini (2013) points to its relationship with resilience in postdisaster reconstruction as pertaining to social ties and networks. Others such as Unwin (2016) have similarly suggested that social capital works at different levels including communities, organizations, and institutions. It is clear from many studies (Aldrich, 2012) that communities with a high reservoir of social capital are able to achieve effective post-disaster reconstruction and build resilience. Often communities that are poor can have great social capital, as pointed out by Aldrich (2012): "...even highly damaged communities with low income and little aid benefit from denser social networks and tighter bonds with relatives, neighbors and extralocal acquaintances." Housing reconstruction can be a vehicle for vitalizing social capital and is an opportunity to form or strengthen community-based organizations that contribute to long-term resilience.

A CHANGING CONTEXT: CLIMATE CHANGE AND URBANIZATION

Scientific evidence indicates the increased frequency and intensity of disasters throughout the world resulting from climate change (Anderson & Bausch, 2006; IPCC, 2012). This has ushered in a new set of challenges: Areas that have historically not experienced certain types of disasters are now experiencing them, such as the floods in Pakistan in 2010 (Gronewold & Climatewire, 2010); disasters are increasingly becoming more frequent and intense in historically disaster-prone areas as in the lower Mekong Delta in Vietnam (Bird, 2009; Vinh, 2012) and the Ganges-Brahmaputra Delta in Bangladesh (Shamsuddoha & Chowdhury, 2007); the entire Pacific region is now subject to loss of land and habitat due to sea level rise, presenting tremendous challenges to human habitat, particularly when compounded by coastal disasters such as cyclones and tsunamis (World Bank, 2013; World Bank & SOPAC, 2009).

Rapid unplanned urbanization is another phenomenon that has resulted in unpredictable disaster patterns. For example, the brisk and often hasty urban development of Bangkok and its surrounding areas in Thailand has subjected it to severe flood impacts as experienced in 2011; as succinctly observed by a United Nations official, "We have grown fast, but not safe" (Barta, 2011). Such urban vulnerability has become strongly evident in a number of developing world cities; the devastation of the 2015 earthquake in Nepal's cities has been pointed out as a consequence of "haphazard urbanization and rampant building code violations" (Misra, 2015) and "increased urban densification, rapidly expanding informal settlements and development that outstrips a government's ability to enforce standards" (Cross, 2015) (see Fig. 12.2). Reconstruction and building resilient housing are thus confronted

FIGURE 12.2

View from Kathmandu, Nepal, in 2009 before the 2015 earthquake showing the unplanned nature of urbanization there.

by a specific set of challenges in cities in developing countries that have densified and grown in unregulated patterns (Habitat for Humanity, 2012; IASC, 2010; IFRC, 2010b). This is discussed further in this chapter in the case study (see Case Study: Villa Rosa, Haiti section).

CHALLENGES IN POSTDISASTER RECONSTRUCTION

While postdisaster reconstruction presents the opportunity for building resilient housing, it often faces a complex range of challenges. Among them stands out the urgency to rebuild houses so that displaced disaster victims have homes again. Reconstruction agencies face pressure from governments, media, and affected communities to build a large number of houses within as little time as possible, as described in the case of Sri Lanka after the 2004 Indian Ocean tsunami (Hettige, 2007; Mulligan, Ahmed, Shaw, Mercer, & Nadarajah, 2012). This is not an easy task given the constraints of a disrupted context; large-scale permanent housing reconstruction is usually a protracted process (Cosgrave, 2008) even in developed countries (Oxfam, 2005) and in developing countries, due to existing institutional and economic shortfalls, attempting rapid reconstruction ushers in a whole set of problems related to institutional arrangements, buildings materials procurement, builder and labor availability, as well as endemic constraints such as corruption. Despite well-meaning intentions, reconstruction agencies are often challenged by obstacles. Reconstruction often requires careful planning, and sensitivity and understanding of the needs of affected communities, which requires time. Mulligan et al. (2012), again in posttsunami Sri Lanka, have argued for such an approach that supports community development beyond simply rebuilding houses.

Housing reconstruction requires addressing a range of issues, each with its own set of challenges: building housing that is culturally and environmentally appropriate; coordination between different agencies and stakeholders; clear policy direction; avoidance of re-creation of vulnerability; equitable distribution; prevention of human rights abuse, corruption, and misappropriation; overcoming inordinate construction delays; and adequate financial management. Such challenges have been discussed widely in the literature (AFP, 2009; Ahmed, 2011; Boen, 2006; Charlesworth & Ahmed, 2015; Eye on Aceh, 2006; Forbes, 2006; INFORM, 2005; Perlez, 2006; Steinberg, 2007; Tsunami Evaluation Coalition, 2006).

THE NEED TO IDENTIFY GOOD PRACTICE

Despite the developing world experiencing the greatest disaster impacts and the housing sector often being most severely affected, there is limited or scattered literature on leading practices on building disaster resilient housing therein. The gray literature of humanitarian agencies often deals with temporary and transitional housing in the context of crises because of their primary focus on emergency management, and literature on permanent housing is comparatively scanty. Most people expect to live in permanent housing over the long term that protects them from multiple disaster cycles, hence the significance of its disaster resilience. The annual "Shelter Projects" reports by UNHCR, IFRC and UN-Habitat (2013–14, for example) is one key attempt to compile examples of housing projects, but the projects are not discussed in much detail. In the face of the limited literature on permanent housing reconstruction, some recent publications such as by Aquilino (2011) and Charlesworth and Ahmed (2015) present some examples of good practice from global case studies. The author of this chapter is a coauthor of the latter publication and has extensively explored housing reconstruction projects around the world. One of the case studies of the book is summarized below as an example of good practice in one of the poorest countries of the world, Haiti, which had been severely impacted by an earthquake in 2010.[3]

CASE STUDY: VILLA ROSA, HAITI

Haiti is one of the poorest countries in the world (Global Finance, 2013) and also highly prone to disaster. The country was struck by a massive earthquake in 2010, which killed 220,000 people and injured more than 300,000. Almost 200,000 houses were badly damaged and more than 100,000 were destroyed. In Port-au-Prince, the

[3]The information presented in the case study has been collected from interviews and observations by the author during fieldwork in Haiti. Some unpublished institutional documents were also consulted.

FIGURE 12.3

A view of Villa Rosa.

capital city, 90% of buildings were informally built and more than 85% of the population were living in slums in tightly packed, poorly built buildings (Blaranova & Christiaens, 2012; United Nations, 2012). Housing reconstruction in Haiti after the 2010 earthquake required addressing a range of social, economic, and other issues, beyond the building of houses. This case study project illustrates such an approach.

The project was located in a settlement in Port-au-Prince called Villa Rosa (see Fig. 12.3), which had grown informally over three decades and before the earthquake had more than 10,000 residents. Typical of such an informal settlement, it was densely built with lot sizes as small as 10 sq. m and narrow pedestrian walkways with limited vehicular access; buildings up to four stories were precariously built on hill slopes (AFH, 2012). The buildings were made of concrete block masonry confined within a reinforced concrete frame and used poor-quality materials. Infrastructure and services were generally lacking. Poor drainage, sanitation, and waste management posed serious health hazards. Sixty percent of Villa Rosa was devastated in the earthquake, and out of 1335 houses, 595 were completely destroyed and 260 damaged.

PROJECT SUMMARY

The Villa Rosa project utilized postdisaster reconstruction as an opportunity to develop and implement a community-based redevelopment and housing plan. It offers useful lessons on how such housing reconstruction can be undertaken in situ in a densely built, urban, poor, informal settlement by working with the disaster-affected community.

Led by the Dutch agency Cordaid and implemented with a range of partners, the Villa Rosa reconstruction project is an example of community-based owner-driven reconstruction together with the development of local skills and capacity, particularly in disaster resilient construction. More than 170 new houses were built and nearly 350 houses were retrofitted with earthquake-resistant construction. The project is also notable for its integration of housing with community infrastructure. This was a pilot project representing the first stage of a long-term community action plan for physical and economic development of the wider Villa Rosa settlement.

AGENCY ROLES

Cordaid provided funding and coordinated with partner agencies, and selected beneficiaries, together with building capacity of the community leaders and community-based organizations. Build Change, a US-based organization specializing in earthquake-resistant construction, supported the physical rebuilding and repair. It also provided training and raised awareness of local builders and homeowners on earthquake-resistant construction, developed house designs through beneficiary consultations, and provided supervision and construction management support. The organization Architecture for Humanity (AFH), supported by UN-Habitat, carried out the physical planning and construction of community infrastructure.

Other agency partners included the International Organization for Migration (IOM) supporting WaSH (water, sanitation, and health); Global Communities (formerly CHF International) ran a cash-for-work program for debris removal; the local authority, Casek, helped regularize tenure security in this informal settlement where previously people did not have land titles. A range of other agencies were involved in the project in smaller and varying roles, indicating the complexity of such projects where it is not feasible, if not impossible, to be implemented by any one agency.

HOUSING RECONSTRUCTION PROCESS

The housing reconstruction process was led by Build Change, initiated in 2011 after clearance of the rubble from the earthquake. In consultation with beneficiaries, Build Change engineers and architects developed the designs for housing and retrofitting. Together with the designs, the first 60% of the funding was provided to the beneficiaries, who then employed local builders trained by Build Change. Construction was supervised by Build Change staff, with a remaining 35% funding provided to the beneficiaries upon satisfactory progress and the final 5% after completion.

For new houses, the US $ 3500 provided was often not enough to complete construction, so beneficiaries added their own funds. Usually the final works that beneficiaries paid for were plastering and painting, which were undertaken after occupying the house when they had available household funds. Two-storey houses were built on some of the smallest lots because of the space limitation. All new and retrofitted single-storey houses included provision for building an extra story in the future.

MAIN PROJECT ELEMENTS

The project consisted of interventions for both physical and economic recovery, comprising the following elements:

- Resilient construction: Earthquake and cyclone resistant construction techniques were applied in all new and retrofitted houses. This included strengthening the building structure and its parts with steel reinforcements, bracings, and straps. A ring beam, lacking in previous houses, was added to hold the structure together. Good-quality thick corrugated iron sheet was used for the roof cladding to avoid lifting off in storms. The process enabled building awareness and local capacity on these resilient construction techniques.
- Community infrastructure: The community prioritized infrastructure upgrading options through a community action plan. This included paved walkways, underground drainage, solar-powered streetlights, landscaping of public areas, and a basketball court popular with the community's youth (see Fig. 12.4).
- WaSH: A range of WaSH services were provided, including communal and household latrines, repair of hand pumps, drilling of wells, and solid waste management.
- Cash-for-work: The cash-for-work program to remove the extensive debris left from the earthquake allowed people to earn desperately needed cash and clear space for the reconstruction to proceed.

FIGURE 12.4

The basketball court was an important part of the reconstructed settlement.

KEY ACHIEVEMENTS

The project followed an owner-driven process. Beneficiaries did not merely receive a house but were involved in its design and construction and gained skills in the process. This allowed them satisfaction and to monitor the quality of materials. Technical support by Build Change ensured further quality control. Together with building new houses, retrofitting damaged houses helped cater to a wider group. Disaster resilient features became acceptable by incorporating them into typical Haitian-style houses. The owner-driven process was complemented by in situ reconstruction; instead of relocation, reconstruction in the existing settlement enabled people to continue living where they had existing livelihoods and networks.

An integrated approach was followed where, in addition to housing, basic infrastructure such as paved walkways, drainage, streetlights, waste management, and WaSH were integrated into the project. The roles played by different agencies made this integration possible. Beyond physical elements, social aspects were also integrated, including land tenure, community organizations, and capacity building. There was provision for incremental expansion by incorporating scope for future vertical expansion of houses, which was an effective strategy given the high density of the settlement and its growing population.

Capacity building through training local builders in disaster resilient construction meant that the skills would remain within the community. Additional training to homeowners helped raise awareness on resilience. The teamwork between different agencies, each bringing its own set of specialized capacities and resources, contributed to the project's achievements and effective implementation.

KEY CHALLENGES

While there were many achievements of the project, in such an impoverished country like Haiti, understandably, there were also significant challenges. Coordinating timely and useful contributions from a wide range of stakeholders was difficult. Aligning toward a common goal the diverse interests from agencies with different backgrounds and institutional cultures was a complex challenge. Apart from the involvement of the local authority Casek, the Haitian government was not directly involved with the project, leaving leadership and decision making to external agencies. Nonetheless, Build Change did initiate public sector technical capacity building by training of government engineers, indicating a way forward to ensure sustainability after external agencies depart.

This was a pilot project, but it came with the risk of creating an "island of benefit" in a "sea" of vast deprivation. This is a common dilemma in many such reconstruction projects in the developing world where poverty and need are widespread and extensive; there simply seems to be never enough resources to make a significantly large-scale impact.

Although local capacity building and a cash-for-work program were provided, there was no support for long-term livelihoods and the economic sustainability of the community. The transition from postdisaster reconstruction to sustainable development remains a typically weak area in the work of humanitarian agencies.

BENEFICIARY NARRATIVES

Although there was a strong technical element in the Villa Rosa project, as discussed above, building resilient housing behind the technical inputs was a human interest factor, as became evident from the personalized narratives of a range of stakeholders involved in the process, documented by the author during fieldwork in Haiti. These narratives highlighted the intangible human motivations, experiences, feelings of satisfaction, and aspirations that act as a mesh reinforcing the achievements of the design and construction of resilient housing. Furthermore, they painted a picture of the life of disaster-affected communities in the developing world.

Here two such narratives are summarized, one from a beneficiary of a new house and another from a retrofitted house, representing the two main modes of housing reconstruction support. Such households, the unit of the community, provide an insight into the community's human experience of the housing reconstruction process.

Owner of a New House

Tesie Lems lived with his wife, child, sister, and two brothers in a small two-room single-storey house built through the Villa Rosa project (see Fig. 12.5). He also benefited from the cash-for-work debris removal program. His previous house had completely collapsed in the 2010 earthquake. He and his family lived in a temporary shelter for close to 3 years before his house could be completed in November 2012.

FIGURE 12.5

The new house of Tesie Lems.

The house was built in an owner-driven process by a Build Change–trained builder with a five-member construction team. Tesie and his household members also helped—cleaning the site, carrying materials, fetching water, etc.

There was provision in the house structure for building an upper floor, but Tesie did not have money for that when interviewed in 2013, hence he was compelled to live in an overcrowded situation with six persons fitting into the two small rooms. Although a small latrine had been provided in the project, a kitchen could not be fitted into such a small lot, so the household cooked and ate outside in a narrow space between the house and its neighbor. Despite such difficulties, Tesie was happy to have received the house. “It’s a good project,” he said.

Owner of a Retrofitted House

Venite Clerilus lived with her infirm husband and four grown-up children in a house retrofitted in the Villa Rosa project (see Fig. 12.6). The house had been partially damaged by the earthquake in 2010; its back part had given way and crumpled. She was the family’s main breadwinner and earned a living by buying wholesale small goods from the countryside, such as vegetables and coal, and selling them in the city for a small profit.

FIGURE 12.6

The retrofitted house of Venite Clerilus.

The back part of the house was rebuilt, and corner reinforcements and a ring beam for earthquake resistance were incorporated into the house's structure. To allow building a future upper floor, reinforcement rods are kept extended beyond the roof, and space retained to add a staircase. With her own funds, she fixed her latrine and built a small septic tank.

The US $ 1500 she received from the project helped her to make the repairs and retrofitting, employing workers trained by Build Change. "I wanted to have a concrete roof so that the house can hold up against storms," she said, and with her savings, and some from her children, she had managed to build such a strong roof. She still needed to paint the house and was saving money for that at the time of interview.

LESSONS FROM VILLA ROSA

The Villa Rosa project was multifaceted and implemented in partnership between several agencies, extending beyond only rebuilding houses destroyed by the disaster to a wider set of inputs to build community resilience. It thus offered some valuable lessons.

Despite the challenges involved in coordinating a range of agencies, this project demonstrates that a multistakeholder and multidisciplinary partnership between the funding agency Cordaid, technical agency Build Change, and urban planning agencies AFH and UN-Habitat, as well as various others, was fruitful in dealing with a complex problem. Bringing on board a range of prominent international partners such as IOM and Global Communities, and also the involvement of the local government authority, Casek, maximized the project's potential for effectiveness.

By avoiding the typical "one-size-fits-all" approach on "green field" sites where communities are resettled as often done in reconstruction projects in the developing world, this project's in situ reconstruction approach succeeded in addressing a diverse range of community needs including livelihoods and social capital that would otherwise not have been achieved.

The project underscores the importance of an integrated approach where housing was combined with community infrastructure such as paved walkways, drainage, streetlights, and landscaping, and services such as water supply, latrines, and waste management.

CONCLUSION

Given the specter of increasing disaster impacts in a fragile and turbulent world, this chapter underscores the necessity of prioritizing building resilience, particularly for the world's majority who are mostly at risk. The developing world, representing the vast majority of the world's population, extensively experiences the severe impacts of disasters, which are amplifying with global climate change and rapid, unplanned urbanization. These impacts are particularly harsh on the housing sector, usually the most valuable and important asset people own, or find shelter in, and quite often not built to be able to withstand these impacts. Thus, there is a widespread need for disaster resilient housing in the developing world. However, although "hardening" of the built environment to withstand and easily recover from disaster impacts is an important aspect of resilience, there is a "soft" aspect, relating directly to people and institutions, particularly manifested in livelihood support and development, local capacity building, and strengthening social capital.

The recent discourse on postdisaster reconstruction, particularly after the 2004 Indian Ocean tsunami, one of the most massive disasters in the near past, focuses on "building back better" strategies as a way of achieving resilience and this is now on the radar of humanitarian agencies. However, there are many challenges to that aspiration in the developing world. Once again, it is not only a matter of rebuilding stronger houses, which in itself can be a challenging task, but also addressing a gamut of issues that sustain the initiative over the long run.

Given this situation, identifying examples of good practice can be a useful building block toward more widespread implementation of resilient housing. As described in this chapter, looking at a case study from one of the poorest and severely disaster-affected countries of the world, Haiti, where achievements have been made despite severe obstacles brings replicable lessons. The country already had extensive impoverishment before the 2010 earthquake and had been hammered by a cyclone just a couple of years earlier. The earthquake exacerbated the existing chronic poverty, creating a highly challenging situation. The key elements of the Villa Rosa case study project—resilient construction, in situ reconstruction in a dense urban informal settlement, partnership between diverse agencies, owner-driven reconstruction, and local capacity building—can pose challenges even in less severe circumstances, but yet the project achieved a significant level of success. This project, with its many positive achievements in such a difficult context, is thus an example of effective postdisaster housing reconstruction in the developing world.

REFERENCES

ADB and World Bank. (2010). *Pakistan floods 2010: Preliminary damage and needs assessment*. Manila: ADB (Asian Development Bank) and Washington, DC: World Bank.

ADRC. (2012). *Natural disaster data book 2010: An analytical overview*. Kobe: Asian Disaster Reduction Center (ADRC).

ADRC. (2014). *Natural disaster data book 2013: An analytical overview*. Kobe: Asian Disaster Reduction Center (ADRC).

AFH. (2012). *Community action plan for Villa Rosa and Sainte Marie*. Port-au-Prince: AFH (Architecture for Humanity).

AFP (Agence France-Presse). (December 27, 2009). Sri Lanka tsunami aid 'misappropriated'. *The Age*.

Ahmed, I. (2011). Overview of post-disaster housing reconstruction in developing countries. *Journal of Disaster Resilience in the Built Environment*, *2*(2), 148–164.

Ahmed, I. (2005). *Handbook on design and construction of housing in flood-prone rural areas of Bangladesh*. Bangkok: Asian Disaster Preparedness Center.

Ahmed, I. (September 2016). Developing disaster resilient rural housing in Bangladesh. In *Paper presented at the sustainable futures conference Nairobi*.

Ahmed, I., & Charlesworth, E. (2013). Disaster risk reduction for sustainable development: The case study of Development Workshop France (DWF), Vietnam. In L. Brennen, L. Lukas Parker, T. A. Watne, J. Fien, D. T. Hue, & M. A. Doan (Eds.), *Growing sustainable communities: A development guide for Southeast Asia*. Prahran, VIC: Tilde University Press.

Aldrich, D. (2012). *Building resilience: Social capital in post-disaster recovery*. Chicago: University of Chicago Press.

Anderson, J., & Bausch, C. (2006). *Climate change and natural disasters: Scientific evidence of a possible relation between recent natural disasters and climate change*. Retrieved from http://www.europarl.europa.eu/meetdocs/2004_2009/documents/dv/ieep_cc_natural_disasters_/ieep_cc_natural_disasters_en.pdf.

Aquilino, M. J. (2011). *Beyond shelter: Architecture and human dignity*. New York: Metropolis Books.

Barakat, S. (2003). *Housing reconstruction after conflict and disaster*. London: Humanitarian Practice Network.

Barta, P. (October 22, 2011). *Floodwaters reach Bangkok*. The Wall Street Journal.

Bird, J. (2009). Mekong climate change and adaptation initiative and the Mekong River Commission. In *Presentation at the Mekong Delta climate change Forum, can Tho, Vietnam*.

Blaranova, L., & Christiaens, B. (2012). *Project proposal: Community development Delmas 30*. Port-au-Prince: IFRC (International Federation of Red Cross and Red Crescent Societies).

Boen, T. (2006). Building a safer Aceh, reconstruction of houses, one year after the Dec. 26, 2004 tsunami. In *Paper presented at the 40th anniversary of Trisakti University, Jakarta*.

Charlesworth, E., & Ahmed, I. (2015). *Sustainable housing reconstruction: Designing resilient housing after natural disasters*. London: Routledge.

CMC (Climate and Migration Coalition). (2015). *Understanding a slow disaster: Getting to grips with slow-onset disasters, and what they mean for migration and displacement*. Retrieved from http://climatemigration.org.uk/understanding-a-slow-disaster-getting-to-grips-with-slow-onset-disasters-and-what-they-mean-for-migration-and-displacement/.

Cosgrave, J. (2008). *Responding to earthquakes: Learning from earthquake relief and recovery operations*. London: Active Learning Network for Accountability and Performance in Humanitarian Action.

Cross, R. (2015). *Nepal earthquake: A disaster that shows quakes don't kill people, buildings do*. Retrieved from https://www.theguardian.com/cities/2015/apr/30/nepal-earthquake-disaster-building-collapse-resilience-kathmandu.

Delaney, P., & Shrader, E. (2000). *Gender and post-disaster reconstruction: The case of Hurricane Mitch in Honduras and Nicaragua*. Washington, DC: World Bank.

Disaster Watch. (2006). *Archives – South Asia tsunami 2004*. Retrieved from http://disasterwatch.net/tsunamifacts_archive.html/.

ERRA. (2006). *Guidelines for earthquake resistant construction of non-engineered rural and suburban masonry houses in cement sand mortar in earthquake affected areas*. Pakistan: ERRA (Earthquake Reconstruction & Rehabilitation Authority).

Eye on Aceh. (2006). *A people's agenda? Post-tsunami aid in Aceh*. Indonesia: Eye on Aceh and Australia: Aid Watch.

Feinstein International Center. (2011). *Examining linkages between disaster risk reduction and livelihoods*. Medford, MA, USA: Tufts University.

Forbes, M. (March 09, 2006). Aid agencies 'lied' over tsunami aid. *The Age*.

Global Finance. (2013). *The world's richest and poorest countries*. Retrieved from http://www.gfmag.com/component/content/article/119-economic-data/12529-the-worlds-richest-and-poorest-countries.html#axzz2sW6jNTkP.

Gronewold, N., & Climatewire (2010). *What caused the massive flooding in Pakistan?*. Retrieved from http://www.scientificamerican.com/article.cfm?id=what-caused-the-massive-flooding-in-pakistan.

Habitat for Humanity. (2012). *Build hope: Housing cities after disasters*. Atlanta: Habitat for Humanity.

Hettige, S. T. (2007). *Tsunami recovery in Sri Lanka: Retrospect and prospect*. Colombo: SPARC and ActionAid International.

HFHA. (n.d.a). *Housing & livelihoods (factsheet)*. Parramatta, NSW: HFHA (Habitat for Humanity Australia).

HFHA. (n.d.b). *Housing & health (factsheet)*. Parramatta, NSW: HFHA (Habitat for Humanity Australia).

Hillier, D., & Nightingale, K. (2013). *How disasters disrupt development*. Oxford: Oxfam GB.

IASC. (2010). *Meeting humanitarian challenges in urban areas*. Geneva: IASC (Inter-Agency Standing Committee).

IFRC. (2010a). *Final report: Cyclone Aila, Bangladlesh*. Geneva: IFRC (International Federation of Red Cross & Red Crescent Societies).

IFRC. (2010b). *World disasters report: Focus on urban risk*. Geneva: IFRC (International Federation of Red Cross & Red Crescent Societies).

INFORM. (2005). *Update: Human rights issues in the post-tsunami conflict*. Colombo: INFORM Human Rights Documentation Centre.

IPCC (Intergovernmental Panel on Climate Change). (2012). *Managing the risks of extreme events and disasters to advance climate change adaptation*. Cambridge, UK: Cambridge University Press.

Lang, H. (2008). Community housing in post disaster area on Nias Islands, Indonesia: Responding to community needs. In *Proceedings of the 4th international i-rec conference, Christchurch*.

Lucini, B. (2013). Social capital and sociological resilience in megacities context. *International Journal of Disaster Resilience in the Built Environment*, *4*(1), 58–71.

Lyons, M. (2009). Building back better: The large-scale impact of small-scale approaches to reconstruction. *World Development*, *37*(2), 385–398.

Lyons, M., Schilderman, T., & Boano, C. (Eds.). (2010). *Building back better: Delivering people-centred housing reconstruction at scale*. Rugby, UK: Practical Action Publishing.

Marti, R. Z. (2005). *The 2004 hurricanes in the Caribbean and the tsunami in the Indian Ocean*. New York: United Nations Publications.

Misra, T. (2015). *How urban planning failed Kathmandu*. Retrieved from http://www.citylab.com/housing/2015/04/how-urban-planning-failed-kathmandu/391499/.

Mulligan, M., Ahmed, I., Shaw, J., Mercer, D., & Nadarajah, Y. (2012). Lessons for long term social recovery following 2004 tsunami: Community, livelihoods, tourism and housing. *Environmental Hazards*, *11*(1), 38–51.

NHDA (National Housing Development Authority). (2005). *Guidelines for housing development in coastal Sri Lanka*. Colombo: Ministry of Housing and Construction.

Oxfam. (2005). *A place to stay, a place to live: Challenges in providing shelter in India, Indonesia and Sri Lanka after the tsunami*. Oxford: Oxfam GB.

Pacific Disaster Center. (2013). *Number of damaged houses: Typhoon Haiyan (Yolanda)*. Retrieved from http://reliefweb.int/map/philippines/philippines-number-damaged-houses-typhoon-haiyan-yolanda-25-november-2013.

Perlez, J. (July 27, 2006). *Aid groups are criticized over tsunami reconstruction*. The New York Times.

Rockefeller Foundation. (2016). *Resilience*. Retrieved from https://www.rockefellerfoundation.org/our-work/topics/resilience/.

Shamsuddoha, M., & Chowdhury, R. K. (2007). *Climate change impact and disaster vulnerabilities in the coastal areas of Bangladesh*. Dhaka: COAST Trust.

Skinner, R. (1991). Peru: Low-income housing. *Mimar*, *38*, 52–55.

Steinberg, F. (2007). Housing reconstruction and rehabilitation in Aceh and Nias, Indonesia—Rebuilding lives. *Habitat International*, *31*(1), 150–166.

Tro, P. T. T. (2011). *Atlas/compendium of housing vulnerability and measures for safe housing in Vietnam*. Hue, Vietnam: Development Workshop France.

Tsunami Evaluation Coalition. (2006). *Joint evaluation of the international response to the Indian Ocean tsunami: Synthesis report*. London: Tsunami Evaluation Coalition.

UNHCR, IFRC & UN-Habitat. (2013–2014). *Shelter projects*. Geneva: UNHCR (United Nations High Commission for Refugees), IFRC (International Federation of Red Cross and Red Crescent Societies) and UN-Habitat (United Nations Centre for Human Settlements).

UNISDR. (2009). *Terminology on disaster risk reduction*. Geneva: UNISDR (United Nations Office for Disaster Risk Reduction.

United Nations. (2012). *Key statistics: Haiti relief*. New York: UN Office of the Secretary-General's Special Adviser.

UNOCHA. (2015). *Nepal: Earthquake 2015 situation report no. 20*. Kathmandu: UNOCHA (United Nations Office for the Coordination of Humanitarian Affairs) and Office of the Resident and Humanitarian Coordinator in Nepal.

UNOCHA. (2016). *Humanitarian bulletin: Cyclone Winston – Fiji*. Suva: UNOCHA (United Nations Office for the Coordination of Humanitarian Affairs) Regional Office for the Pacific.

Unwin, J. (2016). *Why we need to build social capital in cities*. Retrieved from https://www.jrf.org.uk/why-we-need-build-social-capital-cities?utm_content=buffera69d1&utm_medium=social&utm_source=twitter.com&utm_campaign=buffer.

Valdes, H. M., Amaratunga, D., & Haigh, R. (2013). Making cities resilient: From awareness to implementation. *Journal of Disaster Resilience in the Built Environment*, *4*(1), 5–8.

Vinh, K. Q. (2012). Climate change in can Tho and Mekong Delta: Status and ability to cope. In *Presentation at the 3rd high level seminar on environmentally sustainable cities, Siem Reap, Cambodia*.

Wilderspin, I., Barnham, J., Gill, G., & Lockwood, H. (2008). *Mainstreaming disaster risk reduction into DG ECHO's humanitarian actions: Evaluation report*. Essex, UK: Aguaconsult.

World Bank & SOPAC. (2009). *Preparedness, planning, and prevention: Assessment of national and regional efforts to reduce natural disaster and climate change risks in the Pacific*. Washington, DC: World Bank and Suva: SOPAC (South Pacific Applied Geoscience Commission).

World Bank. (2013). *Acting on climate change & disaster risk for the Pacific*. Washington, DC: World Bank.

CHAPTER

Ten Years of Great Love—The Evolution of Housing Reconstruction in Banda Aceh, Indonesia

13

David O'Brien[1], Catherine Elliott[1], Brendon McNiven[2]
[1]The University of Melbourne, Melbourne, VIC, Australia; [2]Arup International, Melbourne, VIC, Australia

INTRODUCTION

The scale of devastation caused by an earthquake in the Indian Ocean on December 26, 2004, was unprecedented. The earthquake, 150 km from the Acehnese coast, affected more than a dozen countries from South East Asia through to East Africa. The energy released on the Earth's surface was more than 1500 times that of the Hiroshima atomic bomb. The seabed rose 5 m in some areas and displaced 30 km^3 of water. Between 9.1 and 9.3 on the Richter scale, the earthquake was the second largest ever recorded and took place over a 10-min period.

It took less than an hour for the first tsunami to reach Aceh's coastline with waves reaching more than 7 km inland (Eye on Aceh, 2006). Of the seven most devastating waves, the maximum was 24 m at the shoreline rising to 30 m inland. A 20-m flood remained temporarily over coastal areas with sand deposits of up to 800 mm thick (Moore, 2006).

Livelihoods were destroyed, livestock was lost, agricultural land was covered in debris, and salt and water sources were contaminated (Eye on Aceh, 2006). The impacts on the natural environment and ecosystems were significant and widespread—affecting agriculture, aquaculture, and fisheries—and local economies were devastated with the inundation of large tracts of coastal land (United Nations Environment Programme, 2005). More than 1 million survivors fled to refugee camps or were housed by friends and relatives (Eye on Aceh, 2006). The World Bank estimated damage and economic loss amounting to US$4.45 billion, almost 97% of the gross domestic product of Aceh (Nazara & Resosudarmo, 2007; United Nations Development Programme, 2005). A pledge of US$7.8 billion was made by over 600 agencies including governments, humanitarian aid agencies, and philanthropists throughout the world. By 2009 US$7.41 billion was disbursed (Aceh Recovery Newsletter, 2009).

This chapter examines two new housing settlements constructed by Tzu Chi, a prominent international aid agency, in Panteriek and Neuheun in the city of Banda

Urban Planning for Disaster Recovery. http://dx.doi.org/10.1016/B978-0-12-804276-2.00013-X

Aceh. These were called "Great Love Village I" and "Great Love Village II," respectively, and were some of the largest reconstruction programs operating in Aceh province with 716 houses constructed at Panteriek and 846 at Neuheun. Many Acehnese were eager to move to these houses having lived in tents and barracks for up to 2 years.

In many respects these two "Great Love" villages are similar in their spatial layouts and identical in their house designs. In both locations the agency used contractors to design and build "duplex" houses of 36 m^2 using the same plan and the same lightweight materials. The most significant difference stems from their locations in relation to the city center. While Panteriek is located 2 km from the city center and is well served by public transport, the Neuheun village is 17 km away, with poor transportation options.

The construction systems used in these houses have not proven to meet the needs of all the residents, and there is some limited awareness and disquiet among some residents regarding the use of asbestos as a construction material rather than the more desirable masonry construction methods. This has contributed to ongoing efforts by a sector of households within these communities to modify and rebuild in an effort to improve their housing. Given that these two villages are so similar in their formal layout, an analysis of both communities allows us to record these modifications and rebuilding efforts and report upon the role location plays in determining the resident's capacity for recovery.

RECONSTRUCTION PROCESSES AND OUTCOMES

More than 200 humanitarian agencies worked in Aceh province after the 2004 Indian Ocean tsunami with 60% engaged in housing reconstruction efforts (Steinberg, 2007). The process adopted three key stages. During the initial emergency response, agencies, where possible, provided tents on cleared land between the destroyed houses. Alternatively people were accommodated in public buildings, such as mosques and schools, in locations inland away from the floods. Many others found shelter with host families or relocated to other parts of Indonesia.

The need for more durable and sturdy shelter types became pressing, and during the second reconstruction phase many agencies began constructing "transitional" housing of a standard between a tent and a permanent house. Two main types of transitional houses were built. Houses of the first type were timber "barracks" shared by multiple families or individuals and located on public land. The second-type houses, supplied by the Red Cross, were 36 m^2 "shelters," steel framed with plywood panels for walling and flooring. After the first monsoon, some reconstruction agencies responded by improving some shelters with additions such as concrete flooring and low brick walls to protect from flooding and rainwater (Oxfam, 2005).

The third key phase, the construction of permanent housing, proved to be a protracted process subject to many technical, social, and governance issues. The initial tsunami response was carried out by the National Development Planning Agency (BAPPENAS) together with the Ministry of Public Works. Within 4 months a separate agency, *Badan Rehabilitasi dan Rekonstruksi* (Agency for Rehabilitation and Reconstruction, BRR), was established. Initially BRR took on a supervisory role, but by the end of 2005 it expanded to 500 staff plus consultants to manage all aspects of the recovery process and implement longer-term projects (Nazara & Resosudarmo, 2007; Steinberg, 2007). Mandated to operate for 4 years, BRR ceased operation in April 2009 with projects worth US$800 million remaining incomplete. Responsibility was delegated to the Ministry of Public Works and coordinated by new agency, the Aceh Sustainable Reconstruction Agency (BKRA) (Aceh Recovery Newsletter, 2009).

Irrespective of the type and size of house owned before the tsunami, BAPPENAS revealed its policy to ensure that survivors were treated equally by reconstruction agencies allocating houses. BAPPENAS stipulated that each reconstruction house should have a maximum floor area of 36 m^2 (Steinberg, 2007). This modest constraint enabled the reconstruction agencies to provide a large range of housing types from a variety of construction materials and to varying degrees of finish and quality. Over the next 4 years 141,000 permanent houses were constructed in the third phase of the reconstruction process. As the redevelopment process unfolded, it became apparent that participating agencies concentrated their efforts in different ways. Some focused on quantity, whereas others built fewer houses but with quality materials and higher levels of finish. Others used the opportunity to shape a new community with social, educational, and religious purposes. A small number of agencies ignored guidelines altogether to compete for influence and prestige by building larger houses with expensive finishes (Das, 2007; Greenblott, 2007; Steinberg, 2007) and many Acehnese comment that some types of houses are far more desirable than the more modest types. However, the typical house included two bedrooms, a living room, a kitchen, and a toilet.

Comparisons of physical outcomes and community attitudes to the various types of houses constructed by the agencies have proven to be revealing. A close study of nine settlements by O'Brien and Ahmed (2012a) has revealed the following series of key issues that affect people residing in the houses built by reconstruction agencies in Aceh.

Residents living in houses built by reconstruction agencies frequently aspire to:

- enlarge the overall size of their house;
- maximize the economic capital of their house;
- demonstrate their status;
- create space for social networking to occur; and
- improve the functionality of their house.

The following are the additional points of interest:

- Around 95% of residents modify their houses.
- Resident's housing aspirations are not met with the types of housing built by the reconstruction agencies.
- The size of the resident's plot of land plays a significant role in defining the options available for modifications.

- The reconstruction process could be streamlined if the designers preplanned for various types of modifications.
- Many residents find the rebuilding process challenging because it involves economic, social, and cultural dimensions.
- A range of house types and sizes is beneficial, as this will accommodate the variety of needs (O'Brien & Ahmed, 2012a).

It is a common theme to see that many households have found the reconstruction agency housing to be insufficient for their needs, and there is significant evidence showing that households are modifying their housing, if and when they have the opportunity, due to factors driven by aspirations with both local and global dimensions (O'Brien & Ahmed, 2014). Many of these modifications are in response to the perceived inadequacies of the house and the inability of the house to meet the needs and aspirations of its residents. For example, it is not surprising that residents with larger families might want to increase the size of their house. Nor is it surprising to find that people who run a home-based business might like to open a shopfront. There are many examples where pragmatic reasons such as these have governed the modification of reconstruction types. In many other cases the residents have sought to match the levels of refinement seen on the better quality houses and those with higher levels of status.

This chapter takes a slightly different angle and compares the same type of house constructed in two different locations by the same reconstruction agency. These two settlements were built using a "cookie cutter" approach building duplex housing with standardized plans and materials. This comparative survey isolates spatial, formal, and technological issues and allows comparisons to be made based on geography and occupant demographics.

TZU CHI IN BANDA ACEH—TWO SITES

Tzu Chi has achieved much renown for its programs over many decades. With humble beginnings from 1966 a Buddhist nun, Dharma Master Cheng Yen, has led the group to be one of the most significant not-for-profit organizations operating internationally. The organization has been built upon Buddhist philosophies and Cheng Yen's works with the poor in the Hualien region of Taiwan (O'Neill, 2010). Initially set up as a collective of housewives supporting the disadvantaged in the local neighborhood, the capacity of the networks grew with help from the media and support from politicians. At the same time an extensive grassroots supporter base of volunteers, many recruited in their time at school, helped implement and finance its charitable programs (O'Neill, 2010).

Tzu Chi now claims more than 10 million members across 47 countries with its devotees contributing to projects that include the development and ownership of several large hospitals, schools, televisions stations, housing estates, and industrial complexes. While Tzu Chi has been highly visible during international disasters and recognized for contributions after Hurricane Katrina (2005), the Indian Ocean tsunami (2004), and the Sichuan earthquake (2008), it has also increasingly been subject to both domestic and international scrutiny. Allegations have been aired in Taiwanese and Chinese media accusing Tzu Chi of using donated funds to purchase high-risk shares in multinational companies (Kuo-tsai, 2015), using its media arm to interfere in a tainted food scandal (Chang, 2014), improper land deals (Kuo-tsai, 2015), tax evasion and poor governance (Hsu, 2015), and moving away from its core relief works (Chen, 2015).

As one of the first relief agencies to respond to the earthquake and tsunami that struck the Indian Ocean on Boxing Day 2004, Tzu Chi played a major role in redevelopments in Sri Lanka and Indonesia and maintained its efforts with large-scale programs over the following 4 years. At Hambantota in Sri Lanka, Tzu Chi was responsible for 649 houses with 1002 houses at Meulaboh, 270 km via the coastal road south of Banda Aceh. A further 716 houses were constructed at Panteriek and 846 at Neuheun in Banda Aceh, and it is these second two settlements, located in Fig. 13.1 below, that form the basis of this study. In Aceh province these case study settlements were labeled "Great Love Village I" at Panteriek and "Great Love Village II" at Neuheun. In each location Tzu Chi also constructed a school (prep to year 9), a medical center, and a community building. With infrastructure services such as water, power, and sewerage plus roadways and drainage, these projects were significant undertakings.

FIGURE 13.1

An aerial photograph locating Panteriek close to Banda Aceh city and Neuheun on the outskirts of the city between the coastline and the foot of the mountains.

CONSTRUCTION TECHNOLOGIES AT THE GREAT LOVE VILLAGES

The dominant form of construction for residential dwellings in Aceh prior to the tsunami has been based on reinforced concrete slab and frame elements with infill panels built from rendered masonry. This system is preferred in urban areas as it is perceived as robust and representative of the household's aspirations toward improved amenity and status. In general, the larger number of agencies involved in the reconstruction efforts followed this pattern, but there were exceptions. These include developments by Bank Mandiri and Uplink that used timber frames and walls and the developments by Muslim Aid and Tzu Chi that used lightweight asbestos panels on timber and steel frames, respectively (Fig. 13.2).

The issue of the use of asbestos is not the key focus of this chapter, but its use in the "Great Love" settlements is significant and subject to further research by the authors. Despite the well-documented evidence demonstrating that prolonged inhaling of airborne asbestos fibers is known to cause serious lung illnesses and death, asbestos continues to be mined and produced in a number of countries, including Russia and China. Western countries phased out asbestos mining and the use of asbestos products in the 1980s before moving to ban it altogether (Australian Government, 2015). The asbestos used in the exterior walls of the Tzu Chi housing is known as *Chrysotile* asbestos and is one of the more common forms of asbestos used in building construction. It is known to cause mesothelioma, a form of lung cancer (Kanarek, 2011).

There is evidence that the asbestos is susceptible to damage from wear and tear in the villages constructed by Tzu Chi. Evidence of aging was observed in the form of cracked corners and instances of "lifting" or delamination of the asbestos sheets from the underlying framing. Walls are dusty with the fibers flaking. Interviews undertaken for this research (and discussed later in this chapter) reveal that although many of the occupants of the houses were aware that their houses were constructed from

FIGURE 13.2

Tzu Chi-constructed duplex type houses with asbestos cladding on a light steel frame. The left image shows a typical duplex at Neuheun in 2009 and the right image shows extensive modifications at Panteriek in 2015.

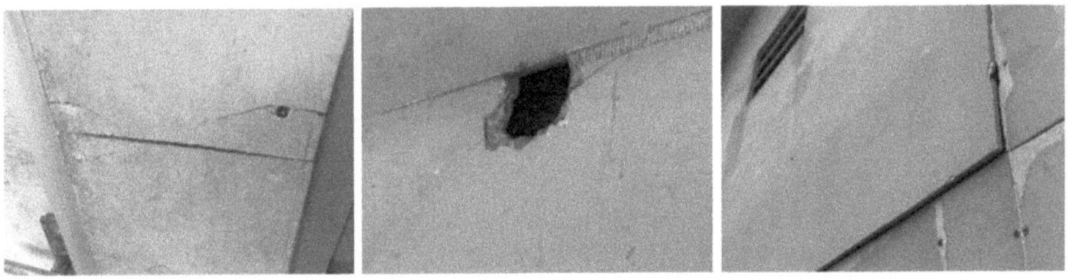

FIGURE 13.3

Examples of cracking and panels lifting in the asbestos sheets in the housing constructed by Tzu Chi.

asbestos, with most complaining about the asbestos dust, few were aware of the health dangers of the airborne particles. Residents were more likely to speak of the limitations of asbestos as a construction material and its lack of robustness by highlighting the cracks and holes in their own houses (Fig. 13.3).

RESEARCH DESIGN

A research team has built a relationship with the community at the Great Love Village II at Neuheun since 2009 revisiting periodically as part of a longitudinal postoccupancy study (O'Brien & Ahmed, 2012a, 2012b, 2014). More recently, the Great Love Village I at Panteriek has been included to extend understandings of the geographic and economic factors that govern residents' modification of houses. A variety of research tactics are required to address questions that have spatial, technical, and social dimensions (Groat & Wang, 2002). To undertake this effectively the research team includes an architect/academic, engineer, and social geographer to analyze data from survey mapping exercises, interviews, and focus groups. Interviews were conducted by Acehnese speaking interpreters to maximize the capacity for interviewees to contribute their experiences.

The team conducted a survey in both the Great Love I (Panteriek) and II (Neuheun) villages to identify a typical street in each and then typical houses in each. The aim was to ensure we included a representative example from each. All houses were mapped and then typical examples (as described in detail later in this chapter) were selected for the interviews. Six groups of interviews were conducted at Panteriek and another six at Neuheun. Additional care was taken when selecting the case study examples to ensure that both genders and a range of age groups were represented.

The research team produced maps, plans, and sketch drawings of case study plots and houses. These were initially hand drawn on site and were later reproduced as a digital database. Data are recorded in a consistent way following typical architectural conventions as shown in Fig. 13.4.

COMPARING VILLAGE MAPS—GREAT LOVE I—PANTERIEK

Fig. 13.5 shows Great Love I Village at Panteriek.

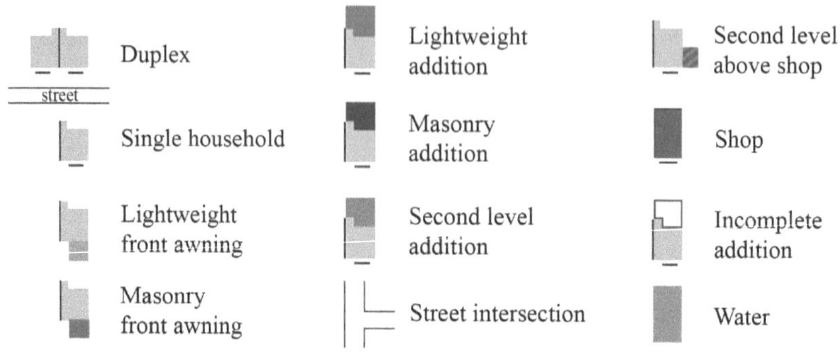

FIGURE 13.4

Key for Figs. 13.5 and 13.6.

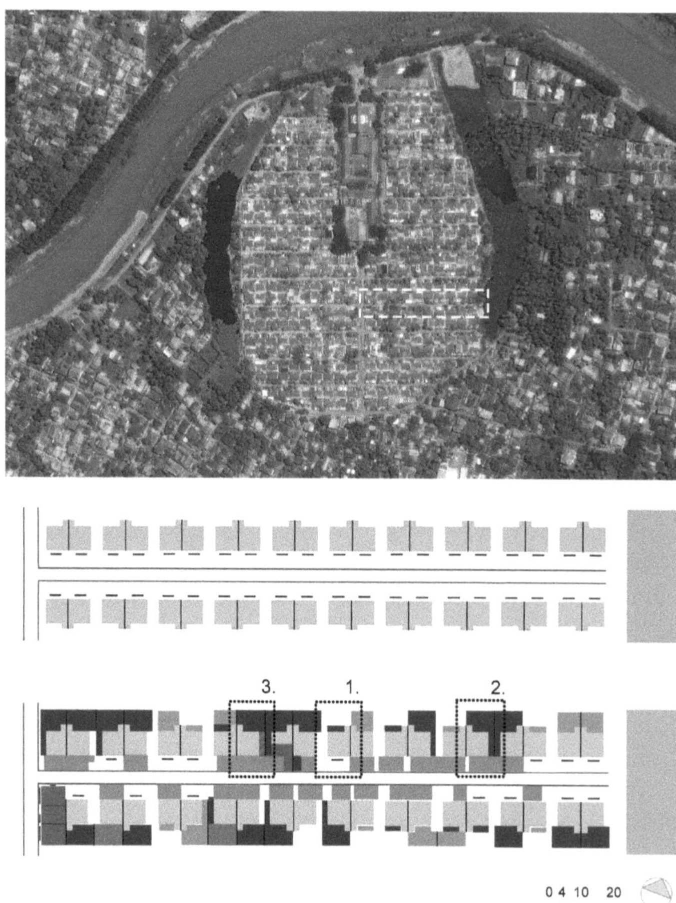

FIGURE 13.5

Great Love I Village at Panteriek. The top image shows an aerial view revealing the layout and scope of the village as it was constructed in 2007 and highlights the location of the street surveyed in 2015 in the drawing below. The case study houses labeled 1, 2, and 3 are discussed in later sections of this chapter.

COMPARING VILLAGE MAPS—GREAT LOVE II—NEUHEUN

Fig. 13.6 shows Great Love II Village at Neuheun.

COMPARATIVE SCOPE OF CHANGE

The maps prepared in the previous section indicate that there is more redevelopment occurring at the Panteriek Village than at Neuhuen. To provide evidence of

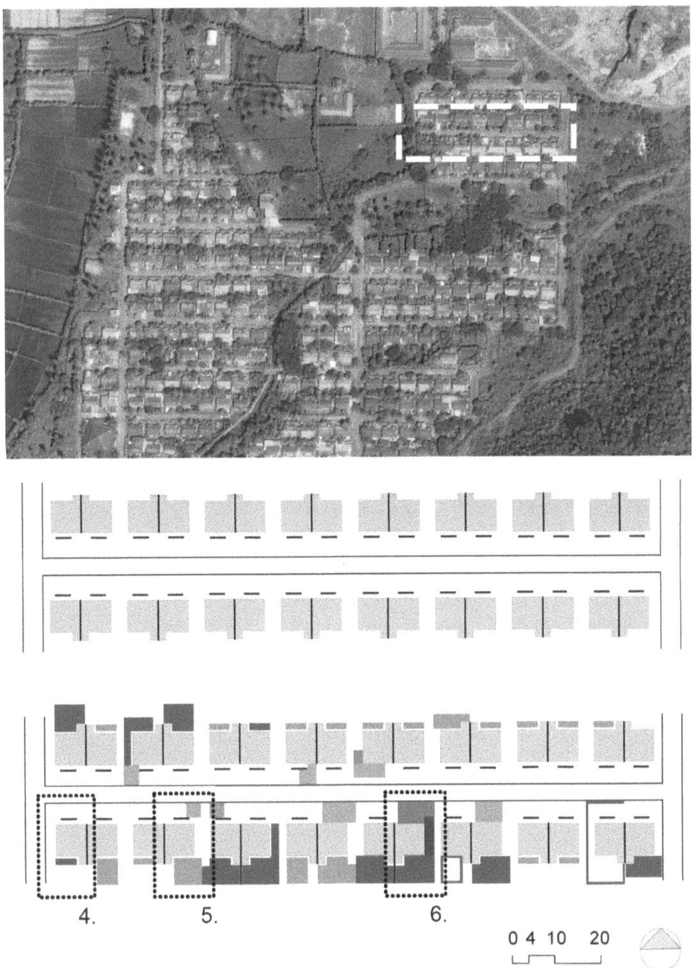

FIGURE 13.6

Great Love II Village at Neuheun. The top image shows an aerial view revealing the layout and scope of the village as it was constructed in 2007 and highlights the location of the street surveyed in 2015 in the drawing below. The case study houses labeled 4, 5, and 6 are discussed in later sections of this chapter.

Level of modification	Type	Great Love Village I Panteriek (32 houses)		Great Love Village II Neuheun (40 houses)	
		Number	%	Number	%
Low-level	A	0	0	1	3
	B	1	3	0	0
	C	2	6	11	28
	D	3	9	2	5
	E	6	19	2	5
Mid-level	F	3	9	3	8
	G	6	19	7 (2 incomplete)	18
	H	4	13	2	5
High-level	I	10	31	3	8
	J	1	3	1	3
	K	2	6	0	0
	L	1	3	0	0
	M	1	3	0	0

FIGURE 13.7

Levels of modification against types at the two Great Love villages.

this, a quantitative comparative analysis identifies relationships between levels of redevelopment across the two settlements. The extent, type, and materiality of the typical levels of redevelopment have been identified—firstly into broad categories (low, mid, and high) and within those escalating levels of improvements (Fig. 13.7).

Low level

Type A: Nil
Type B: Front awning
Type C: Enclosed rear porch in lightweight
Type D: Enclosed rear porch in masonry
Type E: Front awning and enclosed rear porch

Mid level

Type F: Full rear addition in lightweight
Type G: Full rear addition in masonry
Type H: Front awning and full rear additions lightweight
Type I: Front awning and full rear additions masonry

High level

Type J: Front awning, shop, and full rear additions
Type K: Front awning and second storey addition
Type L: Front awning, second storey addition, and shop
Type M: House demolished and replaced by multiple shops

The evidence reveals that while both the Great Love I and II villages have similar quantities of modification, overall there are greater numbers of higher level modifications at Panteriek compared with Neuheun (Fig. 13.8).

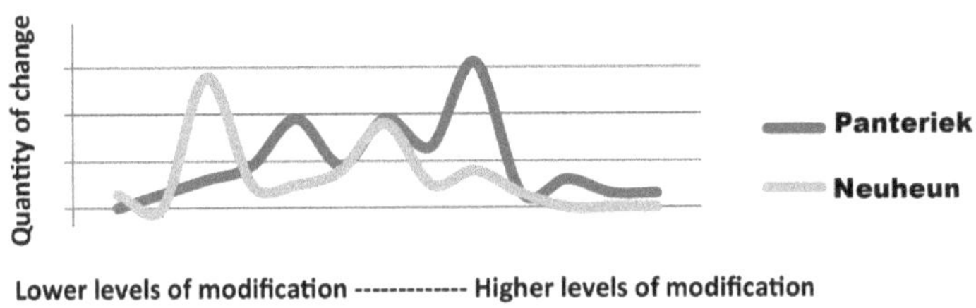

FIGURE 13.8

Comparing levels and quantity of modification at the two Great Love villages.

CASE STUDY EXAMPLES—GREAT LOVE I VILLAGE—PANTERIEK

The case study examples for the three levels of modification are described below (Fig. 13.9).

CASE STUDY 1—LOW-LEVEL MODIFICATIONS

Saufiawati is happy with her house in the Great Love Village I having lived in barracks for nearly 2 years after the tsunami. She particularly appreciates being given a house that is better quality than her house demolished by the tsunami. The appearance of her house is important to her, and she has painted the walls and used filler around the doorframes with the aim to make it appear that the house is not made of asbestos. Saufiawati explained that asbestos is not dangerous in an earthquake, whereas the brick houses crack and fall down. She has added a brick and concrete kitchen at the back, but the foundations are not strong and she thinks it will fall over. Saufiawati has a certificate for the right to use the house (HGB) but is unclear about the formal status (is this in terms of tenure??) of the house and land. She has an HGB for 10 years, but there is no information yet about what will happen after 10 years.

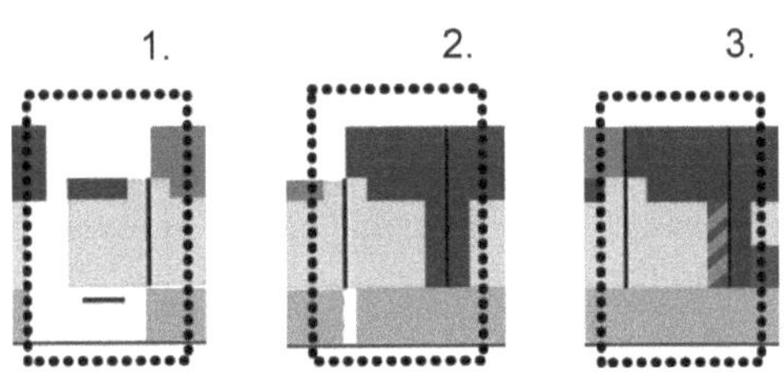

FIGURE 13.9

From left to right, examples of low-level, mid-level, and high-level modifications at Panteriek.

CASE STUDY 2—MID-LEVEL MODIFICATIONS

Nurhayati's tsunami experience has left her too scared to live close to the sea, and she prefers this location since the children can walk to school and the neighbors are nice. She believes that the house is comfortable and with more money plans to make further additions to the house. The layout of the village suits Nurhayati because she can share the cost of extending the house with her neighbors because they can share a wall. She knows that the house is asbestos and is aware of the cracks in the ceilings but says her husband can fix it. She does not believe that there are any serious issues with asbestos because she has lived here for many years and not had any problems. Nurhayati and her husband have a certificate for the right to use the land for 20 years, but she is unsure what will happen then. However, she is confident that they will find a solution in the future.

CASE STUDY 3—HIGH-LEVEL MODIFICATIONS

Sri Iza moved from the coastal area before the tsunami into barracks on the outskirts of Banda Aceh before resettling in her Great Love house in 2007. She now lives with her husband and their three school age children supported by her shop by the house and his work as a driver. The funds for building the shop, and second storey above, came from an Australian nongovernmental organization who had employed her husband. At first the village was not a good place to live and Sri Iza estimates that 200 families allocated houses decided to refuse them because the village was dry, hot, and treeless. She has heard that many went on to regret their decision. She knows that there are better quality houses built by other agencies and sites those built by a Saudi Arabian organization. However, these houses were built much further from town and that is a problem for these families. Despite this she is grateful for the house and sees Panteriek as a good location to earn a living. She has to keep painting the house to try to stop her baby coughing from the dust and the new work on the house has been done in masonry. Sri Iza has a certificate to use the house for 20 years and believes that then she will be given the full certificate.

A common response from the people interviewed at the Great Love I Village was that the location close to the city center offered distinct advantages over many other new villages. This proximity contributed to the capacity of households to build economic security, which in turn, enabled further modifications and improvements to the house. Interviewees had mixed feelings about their tenure over their houses. It was common for some occupants to believe that they were assured 10 years of residency while others thought it was 20. However, this lack of surety makes residents apprehensive about investing in their houses.

In general, the residents are grateful to have been provided with this housing but most are aware that it was not constructed to high standards. Any additional works require better quality foundations, and it is not considered ideal to add to the existing house due to its poor structural capacity. Most interviewees knew their house was constructed from asbestos but remain unsure about the health implications. Symptoms such as coughing were commonly blamed on the asbestos dust, but few knew of any links between asbestos dust and significant health outcomes such as lung cancer.

CASE STUDY EXAMPLES—GREAT LOVE II VILLAGE—NEUHEUN

The case study examples for the three levels of modification are described below (Fig. 13.10).

CASE STUDY 4—LOW-LEVEL MODIFICATIONS

Amina was a tenant before the tsunami and is very grateful to have this house after living in tents and barracks after the tsunami. She is a cleaner and her husband a laborer. They have three children. Amina believes most housing aid was directed to homeowners and she had to work hard to convince five different government agencies to gain permission to register for her house. She understands that her house is not so comfortable because it is dusty. Amina said that the house is far from the city, which is acceptable for the children as they can walk to school. It is more difficult for her to go shopping or to the city as it is a long walk to the bus. Because of this commute, many of her neighbors have left the village to rent new houses closer to the city even though it costs four times as much to rent in the city. Amina spoke of the uncertainty over the ownership certificates and the threat of being asked to leave.

CASE STUDY 5—MID-LEVEL MODIFICATIONS

Suriani lives on her own since her children married and moved to other parts of Aceh. Before the tsunami she was a renter, but in the immediate aftermath she was given a sewing machine and began sewed clothes for her income. She now works making traditional clothes for Acehnese dancing events as well as modern clothes. After living in the barracks in Neuheun for 2 years, she heard that the Great Love II Village would provide houses to renters so she came here. The location of the house is acceptable she says because she is old, she does not have a problem living far away from the city. However, Surianti says the house is low quality because there is asbestos dust, which makes her itchy. She keeps repainting her house every second year to keep the dust in but it does not last.

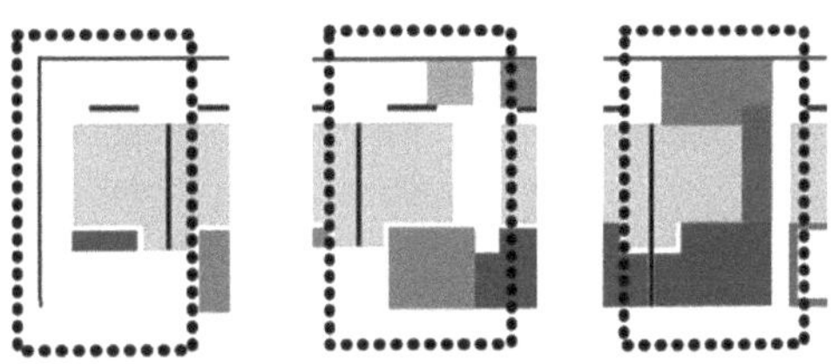

4. 5. 6.

FIGURE 13.10

From left to right, examples of low-level, mid-level, and high-level modifications at Neuheun.

CASE STUDY 6—HIGH-LEVEL MODIFICATIONS

Denny lives with his wife and two children and lived close by before the tsunami. He works as a civil servant in the Department of Fisheries and his house is 16 km from his office. He asked to receive a house from one of the other agencies close to the city but was refused because he had lived in Neuheun before. He then requested to have a house at the Great Love I Village but was not eligible. He has maximized the size of his house by extending to the side and the rear and with a porch to the road. He has used durable materials such as brick and concrete. Denny explains that everyone wants to change to brick rather than asbestos, but the families must work together because of the shared walls. He knows that the house has a problem with the dust and that he should repaint every 2 years. He is aware that there are other panel products that are safer than asbestos, and he is also aware of the ambiguous issues regarding the ownership and the lack of formal certificates. Without an ownership certificate his house is worth far less than others and he cannot get a bank loan.

The people interviewed at the Great Love II Village at Neuheun shared concerns about their long-term ownership over their houses. Efforts have been made by some of the residents to speak with Indonesian government officials and Tzu Chi representatives to clarify the ownership issue but there is no resolution. There is a feeling at this village that the people at the Great Love I Village at Panteriek are more likely to be threatened with eviction because of the higher land value at Panteriek, but no one is really sure of the long-term implications.

Another issue raised by the residents at Neuheun has to do with the lack of a reliable water supply. Residents further to the north of the settlement are elevated above those at the southern end and the water pumps struggle to supply water throughout the day. The residents commented that they could only access water for a short period in the afternoons. At least one resident has attempted to bore his own well, without success, and is frustrated at the lack of service. It is hard to measure the full effect this has upon the residents and their level of investment in their housing, but it is a disincentive for residents to invest in a house that has diminished infrastructure.

DISCUSSION

There are a series of key points emerging from this study that contribute to our understanding of the enabling and disabling factors facing households recovering in the aftermath of a disaster. Housing, for most family groups, ranks alongside health and economic security as a key aspiration within the recovery process and resources are allocated accordingly. Many complex factors govern each household's capacity for recovery, and the research strategy used here highlights the types of common alterations and the role-specific factors such as location, tenure, and materiality play in realizing their ambitions. Tzu Chi's Great Love villages reveal some key points to consider in the light of efforts to improve the resilience of households recovering after disasters.

LOCATION

The evidence reveals that recovery (as measured by a household's capacity to add value to their dwelling) is markedly improved by the location of the settlement in relation to the centers of economic activity. Panteriek, 2 km from the Banda Aceh's city center, demonstrates greater levels of resource allocation to the improvement of the housing as compared with the village 17 km away at Neuheun.

This chapter has identified a few key factors that go toward explaining this difference and each stems from the perceived advantages associated with living closer to the city center. Residents have identified the advantages of good roadways and public transport options that service the Panteriek village. By contrast those at Neuheun describe the difficulties they face commuting from their more isolated village and the extra resources and time this requires. This isolation has other effects upon the economies of the Neuheun village. Whereas the residents at Panteriek have greater capacity to use their house for commercial purposes, given the proximity to higher levels of economic activity, those at Neuhuen have less opportunity. At Panteriek more households have modified their houses to accommodate microbusinesses or shopfronts. An example of this opportunistic initiative is highlighted at Panteriek where a single house at the corner of two streets close to the center of the village has been demolished and replaced with three shops to reap the economic benefits of its corner location.

This inequity is further entrenched during the process of allocating residents to new settlements and houses. Prior to the residents being allocated to specific houses, they were strongly vetted by Tzu Chi staff in a series of interviews. Urbanized homeowners and families with young children were prioritized more highly and allocated houses at Panteriek. Renters, underemployed, and people living on the outskirts of the city were allocated houses at Neuheun. It is important not to overlook the role the redevelopment agency plays as it allocates households to specific locations and houses. In effect Panteriek was the default site for the "elite" clients. Selecting one household for Panteriek and allocating another to Neuheun was a significant piece of social engineering that contributes to differences between the two communities.

There are additional issues at play when the resident's benefit from proximity to the city center. Social networks are nurtured when people live in closer proximity to their shopping, education, health facilities, and family. Panteriek residents find it much easier to travel short distances to the neighboring communities given the quality roads and public transport options. The ease with which the Panteriek residents can engage with their neighborhood is a significant advantage to their well-being and capacity to recover from the disaster. This proximity enhances employment opportunities, which in turn requires less time and resources allocated for commuting to and from work.

TENURE

The residents of both communities stress the importance they place on the issue of tenure and the insecurities they face given the lack of any official certification. The most common belief is that the residents have a 20-year guarantee to occupy the house while some understand this to be only 10 years. No residents have any official government directed explanation of what happens at that time despite many residents seeking further clarification. While this issue clearly does not stop residents from investing in their house, the interviewees mention that this is a risk and that it does influence their decision making. They understand the possibility that they might have to relinquish their houses and be forced to walk away from any investment they have made to the house. However, most families have faith that all will be resolved, "*God willing*." It is not possible to quantify the degree to which this lack of tenure influences resident's behavior; however, the issue must be considered as a factor that dissuades efforts by residents to invest in their housing.

MATERIALITY

The materials used by Tzu Chi to construct the houses, as well as those used in the subsequent additions and modifications, play a significant role in defining the village over the longer term. The robustness of the house, coupled with signs that define the resident's aspirational status, are primary drivers in the reconstruction efforts. The evidence outlined in this chapter points to hierarchies of redevelopment with some materials (asbestos sheet, timber, plywood) seen as less valued as others (masonry, reinforced concrete, tile). This is to be expected in aspirational communities within Aceh (O'Brien & Ahmed, 2014) and more broadly (Hall, 1991; Tagg, 1991).

Tzu Chi's use of asbestos is a key consideration when analyzing the Great Love villages. The interviews with residents touch upon the relationships between asbestos and the perceived lack of robustness, with many comments critical of the brittle material and the ways in which it cracks. Such is the criticism that there are no cases where asbestos is used in any new works initiated by the residents themselves. Instead the research has revealed a preference for timber construction, or for those with the economic means, reinforced concrete and masonry.

As the life span of the asbestos panel diminishes over the years, there is evidence that residents will continue to seek alternate materials as they remodel or demolish houses to rebuild. One house within the sample street at Panteriek has been totally removed, and there is evidence that this process will occur more frequently throughout the whole village. As some residents build higher levels of economic wealth, entire houses across the village are being replaced in a process that is likely to increase—particularly if concerns about the health effects of asbestos fibers are heeded.

DUTY OF CARE

There is a duty of care to not ignore the implications that stem from the use of asbestos paneling within these recovering communities. Although highlighting health concerns was never the focus of this chapter, several questions have been inadvertently raised during the fieldwork period and again during the subsequent analysis. It is worth identifying them here as issues that should be addressed in future research programs. *What responsibilities do reconstruction agencies have over the lifetime of their projects? Does the responsibility end once the structures have been completed or does it extend to include maintenance and harm minimization strategies for building contractors and the broader community? Should Tzu Chi assist and educate the residents of their Great Love villages to identify ways to improve their village without contributing to further undue health risks?* Given the aspirations for improved housing, and the capacity for the residents to drive this change, it is most likely that the trend to improve the housing will continue at a strong pace. A documented and safe technique for removing the asbestos and locating it in secure waste disposal areas must be developed as a duty of care to the residents of the Great Love villages in Indonesia.

CONCLUSIONS

During the reconstruction phase in the aftermath of a disaster, there is significant pressure for international and domestic aid agencies to efficiently produce large numbers of houses to resettle and rehabilitate the affected communities. Limited ranges of designs and mass production techniques were methods employed in Aceh to rapidly increase the quantity of housing. To its credit the Tzu Chi organization was responsible for some of the extensive redevelopment programs in Indonesia in the wake of the 2004 Indian Ocean tsunami. Two of these developments, one close to the city center at Panteriek and the other 17 km away at Neuheun, highlight the importance location and socioeconomic factors play in the redevelopment of the housing infrastructure. The residents in the settlement closer to town (Panteriek) are more likely to be employed and have greater capacity to improve their houses with the addition of more living spaces and improved construction materials. On the other hand, the residents in the more remote settlement (Neuheun) are less likely to have well paid employment and hence have lower capacities to improve their housing. This difference between the two settlements was not accidental with Tzu Chi deliberately selecting urban homeowners to reside at Panteriek and the underemployed or renters to live at the more remote Neuheun village. These factors highlight growing levels of disparity between the two settlements with Panteriek undergoing physical changes, as evidenced by resident-initiated modifications to housing, at a higher pace and exhibiting more substantial levels of remodeling. Other factors governing these differences appear to be of a secondary nature. For example, access to reliable water supplies is less assured at Neuheun, ensuring that the settlement is less attractive to many residents.

These difficulties, coupled with the insecurities that accompany the vague tenure issues, do not stop the majority of residents from both settlements aspiring to make efforts to improve the housing with the addition of new living, sleeping, and gathering places made from more robust construction materials. However, little is known about the best methods for rebuilding (and demolishing) the houses—particularly as asbestos is the main construction material. Given that asbestos fibers are so strongly linked with lung cancer, it is paramount that steps be put into place to educate these communities of the dangers as well as providing residents, and the construction sector, with safe and cost-effective ways to remove and treat the waste asbestos. The role that Tzu Chi might play in future education and rehabilitation programs at Neuheun, Panteriek, and Meulaboh should be amplified given the organization's charter to care for people in need.

ACKNOWLEDGMENTS

The authors would like to thank the people from the Great Love villages at Panteriek and Neuheun for hosting the research team and providing their valuable insights. They also thank Sazali M. Syukur Ismail (Jali) for his significant contribution to the interview components of this project.

REFERENCES

Aceh Recovery Newsletter. (2009). *3*(5) Banda Aceh.

Australian Government. (2015). *Australian Government Department of Health*. Retrieved from http://www.health.gov.au/internet/publications/publishing.nsf/Content/asbestos-toc~asbestos-when-and-where.

Chang, J. (2014). *Tzu Chi organization criticized for favoring Ting Hsin*. The China Post. Retrieved from http://www.chinapost.com.tw/taiwan/national/national-news/2014/10/21/419946/Tzu-Chi.htm.

Chen, C. (2015). *Tzu Chi row reflects three social changes*. Taiwan News. Retrieved from http://www.taiwannews.com.tw/etn/news_content.php?id=2696775.

Das, J. (2007). *Reconstructing responses: The trials and tribulations of World Vision's tsunami shelter program*. Annual Review 2007. Australia: World Vision.

Eye on Aceh. (2006). *A people's agenda? Post-tsunami aid in Aceh*. Australia: Eye on Aceh, Indonesia and Aid Watch.

Greenblott, K. (2007). *Shelter programming: Learning from the Asia tsunami response*. London: World Vision.

Groat, L., & Wang, D. (2002). *Architectural research methods*. New York: John Wiley & Sons.

Hall, S. (1991). The local and the global. In A. D. King (Ed.), *Culture, globalization and the world-system*. Macmillan.

Hsu, E. (2015). *Why Tzu Chi is sparking resentment*. Taiwan News. Retrieved from http://www.taiwannews.com.tw/etn/news_content.php?id=2698712.

Kanarek, M. S. (2011). Mesothelioma from chrysotile asbestos: Update. *Annals of Epidemiology*, *21*(9), 688.

Kuo-tsai. (2015). *Tzu Chi must step into the light*. Taipai Times. Retrieved from http://www.taipeitimes.com/News/editorials/archives/2015/04/02/2003614946.

Moore, A. (2006). *Sedimentary deposits of the 26 December 2004 tsunami on the northwest coast of Aceh, Indonesia*. Retrieved from http://www.terrapub.co.jp/journals/EPS/pdf/2006/5802/58020253.pdf.

Nazara, S., & Resosudarmo, B. (2007). *Aceh-Nias reconstruction and rehabilitation: Progress and challenges at the end of 2006.* Tokyo: Asian Development Bank Institute.

O'Brien, D., & Ahmed, I. (2012a). Donor-driven housing, owner-driven needs: Post-tsunami reconstruction in Aceh, Indonesia. In N. Kaufman (Ed.), *Pressures and distortions – City dwellers as builders and critics* (pp. 237–338). New York, US: Rafael Vinoly Architects.

O'Brien, D., & Ahmed, I. (2012b). Resident initiated modifications to reconstruction housing in Banda Aceh. In E. Blakely, E. Birch, R. Anglin, & H. Hayashi (Eds.), *Managing urban disaster recovery: Policy, planning, concepts and cases* (pp. 163–169). Berkshire, UK: Crisis Response Publications.

O'Brien, D., & Ahmed, I. (2014). Global and regional paradigms of reconstruction housing in Banda Aceh. *Open House Journal Special Edition Post Disaster Housing Reconstruction, 39*(3).

O'Neill, M. (2010). *Tzu Chi: Serving with compassion.* John Wiley & Sons.

Oxfam. (2005). *A place to stay, a place to live: Challenges in providing shelter in India, Indonesia and Sri Lanka after the tsunami.* Oxford: Oxfam.

Steinberg, F. (2007). Housing reconstruction and rehabilitation in Aceh and Nias, Indonesia – Rebuilding lives. *Habitat International, 31*(1), 150–166.

Tagg, J. (1991). Globalization, totalization and the discursive field. In A. D. King (Ed.), *Culture, globalization and the world-system.* Binghamton: The Macmillan Press.

United Nations Environment Programme (UNEP). (2005). *After the tsunami: Rapid environmental assessment.* Nairobi: UNEP.

CHAPTER

Community-Driven Change

14

Kate Cotter
Bushfire Building Council of Australia, Melbourne, VIC, Australia

INTRODUCTION

This chapter includes a personal account of the sterilization of my land and the community advocacy role I undertook to campaign for changes to bushfire planning policy. I also explore the policy development context between 2009 and 2011, impacts of the policy on the broader community, the advocacy process, and the ongoing challenges for property owners, communities, and policy makers.

On February 7, 2009, now known as Black Saturday, 173 lives were lost and 2029 homes were destroyed by bushfires in Victoria, Australia. The Victorian Bushfires Royal Commission (VBRC), conservatively estimated that the bushfires caused losses of AUD $4.4 billion (VBRC, 2010c), approximately 1.4% of Victoria's state gross product (ABS, 2015). The impacts of Black Saturday remain today, survivors mourn the loss of loved ones, and rebuilding lives, homes, and communities continues to be a challenge.

In response to Black Saturday, new statewide bushfire planning regulations called the "Bushfire Management Overlay" (BMO) were implemented on November 18, 2011. The regulations created their own wave of postdisaster trauma. Private land, previously lawfully subdivided, was rendered undevelopable where sites were deemed an "unacceptable risk." This "sterilization" had considerable local impacts, particularly as the government did not provide compensation, despite the VBRC recommendation to do so.

BLACK SATURDAY REBUILDING AND RECOVERY

The Black Saturday bushfires triggered government and community reevaluation of the state's bushfire risk management strategy. The imminent concern was how and where would Black Saturday victims rebuild and reestablish their lives. The statewide strategic issue was whether planning and building policy could prevent life loss from the inevitable future disaster.

Urban Planning for Disaster Recovery. http://dx.doi.org/10.1016/B978-0-12-804276-2.00014-1

In the 48 h following Black Saturday, the then Prime Minister of Australia, Kevin Rudd, said:

> *Hear this from the Government and the Parliament of the nation. Together we will rebuild each of these communities — brick by brick, school by school, community hall by community hall.*
>
> **Hudson (2009)**

Questions arose about whether people in high-risk areas should be permitted to rebuild at all. Although it seemed distasteful to consider permanent displacement of traumatized survivors, potential acquisition of high-risk property was openly discussed among media commentators, the wider community, and later by the VBRC. The following issues were raised:

- Would a buyback scheme include the properties still standing in those townships? If the risk was too high to rebuild new, bushfire-rated homes, would not the risk be far greater for existing houses in the same area that are not built to any bushfire standards?
- Would all existing high risk properties across the state be compulsorily acquired? What parts of Victoria are too high risk for people to live in and visit? How would that affect the future of Victoria?
- Would compulsory acquisition occur as a preventative measure or only apply to property destroyed after disasters?
- What is the threshold for unacceptably high risk?
- Can bushfire risk be addressed through building and landscape design?
- Can the broader landscape be managed to reduce bushfire intensity?
- Is the planning and development strategy either to retreat from bushfire or to accept it and address it?
- Will evacuation policy provide an alternative to relocation and property acquisition?

Many Black Saturday victims felt that rebuilding was essential to their personal recovery process (Gunningham, 2015). Government responded to this sentiment by permitting Black Saturday victims to rebuild on the same sites and granted exemptions from bushfire planning regulations on May 14, 2009 (VC57, 2009). For victims who did not want to rebuild, the government established a voluntary buyback scheme. There were four limiting factors to the buyback scheme: it applied only to properties destroyed on Black Saturday, the property must have been a primary residence, it had to be located within 100 m (328 feet) of significant forest, and the total budget was capped at AUD $50 million (Department of Justice, 2012).

Of the 2029 properties destroyed, only 550 were eligible for the buyback scheme, and 114 properties accepted the buyback (Department of Justice, 2012). That is, 6% of the total properties destroyed, and 21% of those eligible accepted the buyback.

To put this in context, there are approximately 300,000 properties covered by bushfire planning regulation in Victoria (DTPLI, 2014) and 90% of Victoria by area is designated as prone to bushfire (Fig. 14.1) (DELWP, 2016). If 5% of all

FIGURE 14.1

Designated bushfire-prone areas of Victoria are shown in gray.

From DELWP (2016).

bushfire-prone sites were deemed an unacceptable risk, the value of those sites would be approximately AUD $7 billion.

The limitations on eligibility of the buyback scheme demonstrated poor political commitment to a statewide risk management and resettlement strategy. Apart from Black Saturday–affected areas, it remained unclear what the state's intentions were for existing homes and new developments in other high risk areas.

THE VICTORIAN BUSHFIRES ROYAL COMMISSION

The VBRC was established on February 16, 2009, to investigate the causes of, and responses to, the Black Saturday bushfires. The Commission handed down 67 recommendations in July 2010, and in November 2010 a newly elected government promised to implement all recommendations. However, many recommendations have since been abandoned or only partly implemented for various reasons including prohibitive cost (BRCIM, 2012). There were 18 recommendations relating to planning and building reform (VBRC, 2010b), which were largely interdependent, so as to provide the state with a consistent land use strategy for all property types: undeveloped sites, existing homes, new homes, and new and existing buildings for vulnerable uses such as childcare centers, hospitals, nursing homes, schools, and accommodation for disabled people. This chapter argues that planning and building recommendations have been selectively adopted, leading to a failed land use strategy with poor outcomes for landowners and communities.

VBRC recommendation 49(h) proposed that vulnerable use buildings should be subject to bushfire building regulations, which has not subsequently been implemented despite the potential catastrophic outcomes for school children, nursing home residents, hospital patients, and people with disabilities, whom all face barriers to evacuation (PWC, 2012).

Recommendation 53 included the requirement that property owners provide a Bushfire Attack Level (BAL) assessment of the site and the construction standard (if any) of the home, to be included in compulsory property sale disclosure documents. Just as the motor vehicle safety star ratings inform purchasing decisions without guaranteeing the prevention of injury or death (ANCAP, 2016), the VBRC sought to empower property buyers with greater knowledge about a site's bushfire risk and building compliance. It was also intended that the market would reward bushfire resilient properties with higher demand and sales prices. Alarmingly, the recommendation was rejected by the Department of Justice without any evidence-based justification:

> *In relation to the disclosure of a current BAL assessment in a section 32, DOJ has advised that concerns were raised as prospective purchasers may rely solely on this instead of undertaking their own due diligence. The current BAL assessment may give prospective purchasers a false sense of security.*
>
> **BRCIM (2012)**

The BMO was developed in response to VBRC recommendations regarding land use planning, which included the recommendation to limit development on sites deemed an "unacceptable risk." To avoid the "harsh consequences for the landowners concerned if land is sterilized" (VBRC, 2010a), VBRC recommendation 46 proposed a government funded retreat and resettlement policy for existing developments and identified buyback and land swap schemes as solutions (VBRC, 2010a). The VBRC also advised government that bushfire planning policy reforms must "explicitly enable landowners to take reasonable steps to reduce bushfire risk to an acceptable level," ensuring that "acceptable risk is clearly defined" (VBRC, 2010a).

The BMO limited development in high-risk areas, as recommended by the VBRC, but the government did not buyback land that the policy rendered undevelopable or offer land swaps or any form of compensation. There was no definition of "acceptable risk," so it was unknown how landowners could reduce risk to an "acceptable level." Mapping was not publicly available to identify the unacceptably high-risk regions or sites, so landowners, home owners, property purchasers, and communities did not know whether they were subject to land sterilization.

MY STORY

This section of the chapter is my personal account of the impacts of the BMO as an affected landowner and the community advocacy role I undertook to lobby for changes to the policy to enable ourselves and other landowners to build our homes.

BIG HILL

In 2004 my family bought land at Big Hill, a coastal settlement on the Great Ocean Road in South-West Victoria. Our land sits in the treetops with views to the ocean (Fig. 14.2).

Our 3-acre site was subject to bushfire planning regulation, and bushfire resilience was central to the design of our home. We planned to build to the highest bushfire construction standards and install an accredited private bushfire shelter.

The purchase of our land included planning permission to build a home, but we could not afford to build within the regulated 2-year time frame. We would have to reapply for planning permission when we were ready to start the project (Fig. 14.3).

Our neighbor at Big Hill started construction of his home in 2011 and had successfully navigated through the existing bushfire and local planning requirements. Our sites are adjacent and nearly identical in size, shape, aspect, and slope; we therefore hoped to achieve the same result—permission to build on our land (Fig. 14.4).

By mid-2011 we were ready to start the project. We hired expert consultants to prepare the necessary planning application reports: BAL site assessment (which determines the level of bushfire risk and corresponding construction requirements) and ecological, geotechnical, and water treatment studies, and we finalized the house design drawings. Prior to submitting our planning application, the BMO was introduced on November 18, 2011. We had to get a new BAL rating and bushfire planning

FIGURE 14.2

Big Hill, Victoria.

Photo courtesy of Kate Cotter.

FIGURE 14.3

Cotter land at Big Hill.

Photo courtesy of Kate Cotter.

FIGURE 14.4

Neighboring property at Big Hill.

Photo courtesy of Kate Cotter.

FIGURE 14.5

BAL ratings.

From CFA (2012).

report to comply with the new planning regulations. To our horror, our BAL rating changed from medium risk BAL 29 to extreme risk BAL Flame Zone (FZ). Although the national construction standards provided solutions for BAL FZ, the BMO did not allow new BAL FZ development at all (Fig. 14.5).

Although we proposed additional safety measures such as sprinklers and an approved private shelter, there was no solution available for our site. We sought second and third opinions and met with the Country Fire Authority (CFA) on site, but the answer was the same—we could not build on our land.

We spent Christmas 2011 in a depressed state. We stood on our now worthless land and wondered if we should have bought one of the older houses across the street instead. We would have been more exposed to bushfire risk, but at least we could have lived where we wanted to live. Our neighbor could build but we could not. None of it seemed logical, fair, just, or reasonable.

THE LOBBY GROUP

Once the feelings of shock and hopelessness subsided, I decided I had to fight. It was not really a choice, we could not afford to lose the value of our land. In March 2012 I set up a blog, where I wrote about our situation and called for affected landowners to join my lobby group. My strategy was to provide a unified voice for landowners across the state, gather data from landowners to establish an evidence base, research the political policy context, collaborate with experts to develop alternative solutions, and work with media outlets to tell our story to the broader community, with the aim of pressuring government to change the regulations.

The blog was generating interest, and by late 2012 the lobby group had over 3000 members. Despite the large number of members, there was broad consensus that we wanted government to amend the BMO to provide solutions for all sites. We believed that was more achievable than compensation that could amount to billions of dollars and the majority of members did not want to give up their land for any price.

RESPONSE FROM GOVERNMENT

Lobby group members wrote to the Planning Minister to explain their situation and argue for reasonable reforms to the BMO. The response from government and agencies was consistent—the VBRC recommended restricted development

from "The Australian", 7 August 2013

Owners threaten action over fire plan

DAVID GERAGHTY

Paul and Amanda Houghton at their Hepburn Springs property, which they can't build on because of tough fire regulations introduced after Black Saturday

PIA AKERMAN

THE Napthine government is facing the prospect of further legal action over its response to the Black Saturday bushfires, with thousands of angry landowners threatening a class action because of planning changes that have rendered their properties worthless.

The government, which is already embroiled in Victoria's biggest class action from victims of the 2009 bushfires, has come under attack for its implementation of bushfire management overlays as recommended by the bushfires royal commission.

The overlays restrict new developments in areas designated at high risk from bushfire, but a lobby group of 4000 members argues that they have been wrongly implemented in some places.

Group founder Kate Cotter said members in regions ranging from the Otways and Lorne on the state's western side to Gippsland in the east had seen their council rates slashed to zero in recognition that their property had lost value.

"There are some lots at the top of a hill surrounded by national park that you would say would be a silly to place to build . . . but 90 per cent of the people in my group are in an infill situation," she said.

"They're just the last block on the street to be developed. If the state government has decided that some areas are too dangerous to inhabit then they need to clearly identify those towns, let communities know that they are shutting down those areas and buy back every property that has been rendered valueless."

The group has met several times with the Country Fire Authority and the planning department to seek a solution but to no avail, and is consulting lawyers on other moves to force amendments or obtain compensation.

Paul and Amanda Houghton bought a block of land on Hepburn Springs main road in 2010 to develop and add to their existing boutique accommodation holdings in the popular spa town.

They have since learned they would be refused permission for any building because the block is considered in a "flame zone", at high risk of fire because of the publicly owned mineral springs reserve behind it.

"I could understand if we were right in the middle of the forest . . . but we're not," Mr Houghton said. "We're right in the middle of Hepburn village.

The action group will hold a public meeting this weekend.

FIGURE 14.6

Lobby group members Paul and Amanda Houghton from Hepburn Springs.

From The Australian, Akerman, P. (2013), Photo: David Geraghty/Newspix.

and the government would not provide compensation, in other words "tough luck" (Figs. 14.6 and 14.7).

We consulted with bushfire scientists and fire safety engineers and proposed solutions that would strengthen life safety outcomes and allow development for existing subdivisions. I presented to the CFA, Fire Services Commissioner, and Department of Transport, Planning and Local Infrastructure (DTPLI) where the response was the same—there would be no shift in policy.

However, a meeting with the then Red Tape Commissioner, Mr. John Lloyd, fortunately resulted in him taking great interest in the issue. He visited lobby group members on their land and met with bushfire experts and government departments.

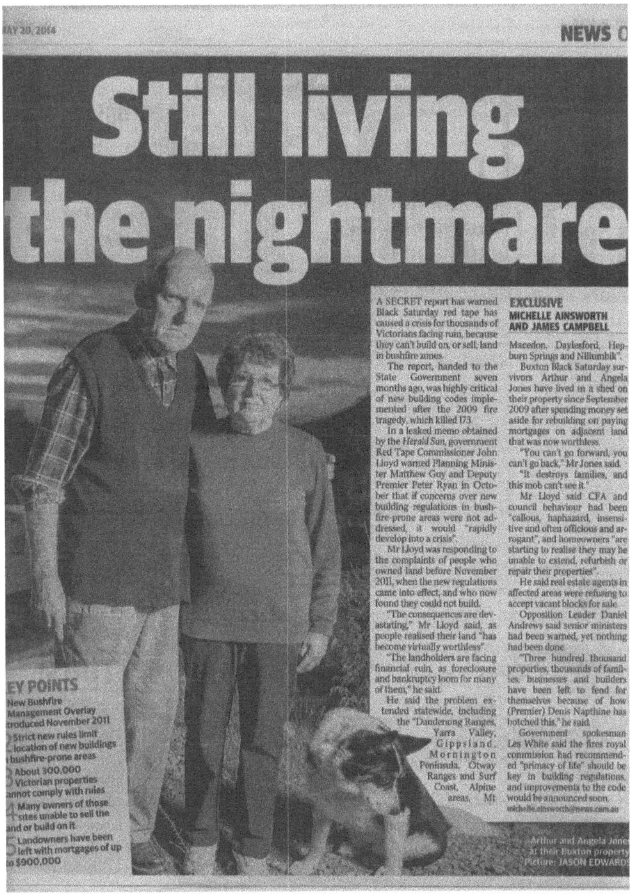

AY 20, 2014 NEWS

Still living the nightmare

EXCLUSIVE
MICHELLE AINSWORTH AND JAMES CAMPBELL

A SECRET report has warned Black Saturday red tape has caused a crisis for thousands of Victorians facing ruin, because they can't build on, or sell, land in bushfire zones.

The report, handed to the State Government seven months ago, was highly critical of new building codes implemented after the 2009 fire tragedy, which killed 173.

In a leaked memo obtained by the *Herald Sun*, government Red Tape Commissioner John Lloyd warned Planning Minister Matthew Guy and Deputy Premier Peter Ryan in October that if concerns over new building regulations in bushfire-prone areas were not addressed, it would "rapidly develop into a crisis".

Mr Lloyd was responding to the complaints of people who owned land before November 2011, when the new regulations came into effect, and who now found they could not build.

"The consequences are devastating," Mr Lloyd said, as people realised their land "has become virtually worthless".

"The landholders are facing financial ruin, as foreclosure and bankruptcy loom for many of them," he said.

He said the problem extended statewide, including the "Dandenong Ranges, Yarra Valley, Gippsland, Mornington Peninsula, Otway Ranges and Surf Coast, Alpine areas, Mt Macedon, Daylesford, Hepburn Springs and Nillumbik".

Buxton Black Saturday survivors Arthur and Angela Jones have lived in a shed on their property since September 2009 after spending money set aside for rebuilding on paying mortgages on adjacent land that was now worthless.

"You can't go forward, you can't go back," Mr Jones said.

"It destroys families, and this mob can't see it."

Mr Lloyd said CFA and council behaviour had been "callous, haphazard, insensitive and often officious and arrogant", and homeowners "are starting to realise they may be unable to extend, refurbish or repair their properties".

He said real estate agents in affected areas were refusing to accept vacant blocks for sale.

Opposition Leader Daniel Andrews said senior ministers had been warned, yet nothing had been done.

"Three hundred thousand properties, thousands of families, businesses and builders have been left to fend for themselves because of how (Premier) Denis Napthine has botched this," he said.

Government spokesman Les White said the fires royal commission had recommended "primacy of life" should be key in building regulations, and improvements to the code would be announced soon.

EY POINTS
New Bushfire Management Overlay troduced November 2011
2 Strict new rules limit location of new buildings bushfire-prone areas
3 About 300,000 Victorian properties annot comply with rules
4 Many owners of those sites unable to sell the and or build on it
5 Landowners have been left with mortgages of up to $900,000

Arthur and Angela Jone at their Buxton property Picture: JASON EDWARD

FIGURE 14.7

Lobby group members Angela and Arthur Jones.

From The Herald Sun, Ainsworth, M. and Campbell, J. (2014), Photo: Jason Edwards/Newspix.

Mr. Lloyd took the case to the Deputy Premier and impressed upon government that the BMO needed to be addressed urgently as it was causing intolerable suffering in the community and would damage the government.

By July 2013 we had not succeeded in engaging with the Minister despite our attempts to produce a "win–win" outcome for government and landowners, through reasonable, evidence-based amendments to the policy. Landowners did not have time on their side, banks were getting nervous about holding mortgages on valueless property, families were spending their building money on rent, and I was receiving phone calls from members who were at breaking point. We were all desperate to get on with our lives and limit the damage already inflicted by the BMO. We launched our media campaign with a view to increasing pressure on government ahead of the 2014 state election.

MEDIA

My media strategy was to present two aspects of the issue, the impacts on landowners and the consequences for regional communities. It was likely that many Victorians would be outraged that government could sterilize land without compensation. It was also likely that most Victorians had a personal connection to the regional townships and settlements that were under threat from the BMO. I worked on developing relationships with media, covering state and national print, city and regional radio, public and commercial television, and freelance journalists. We agreed on communicating the themes of justice, fairness, and reasonable solutions to the media. While I was confident that we could provide a never-ending stream of stories about different families in all areas of the state, I was concerned that media appetite for our stories would be short-lived. Despite my concerns, between July and October 2013 we had coverage across all media outlets on a weekly basis.

THE MINISTER CALLS

On the October 15, 2013, we were featured on a prime-time national current affairs TV program, which resulted in widespread media coverage the following day. The media interest triggered an immediate response from government—I was called in to meet the Minister and his advisors. The Minister said our proposed solutions seemed entirely reasonable and the government would act quickly to alleviate the suffering imposed by the policy. Politicians from both major parties tabled their concerns about the BMO in parliament (Fyffe, 2013, p. 2966 and McLeish, 2013, p. 3619). I realized our campaign was succeeding; we were also receiving support from industry associations, local government, and the broader community.

On December 17, 2013, the Planning Minister announced that reforms would be made to the BMO:

> *I expect those changes will be very straight forward to allow residents to be able to build on their properties, particularly on land that's been deemed unbuildable at this point of time.*
>
> *What we're going to do is ensure private land, private risk. That principle is paramount and importantly Victorians will be aware of their fire risk before they build.*
>
> **Longbottom (2013)**

However, by April 2014 the promised changes to the BMO had not eventuated. We reinstated the media campaign, which led to this newspaper editorial:

> *Premier Denis Napthine is faced with a serious and growing problem as the November 29 election draws closer. The Government cannot stand by as people, through no fault of their own, find themselves paying off loans on land they cannot either build on or sell because of the building codes adopted after the bushfires in 2009.*
>
> **Herald Sun Editorial (2014)**

FIGURE 14.8

Lobby group protest at parliament.

From ABC News (2014).

To maximize the media coverage, we organized a protest at parliament in the following days, where we handed over a petition to the Planning Minister. All TV news outlets covered the story that evening (Fig. 14.8).

The Minister's advisors called me shortly after the protest, and we had several meetings to work through changes to the BMO.

LEGISLATION CHANGES

On July 31, 2014, changes to the BMO planning legislation were implemented. The changes did not go as far as guaranteeing "private land, private risk." The new BMO was not a wholesale change in the way risk was assessed or managed; acceptable risk was not defined and remained open to the interpretation of referral authorities. However, some of our suggested solutions were included; there were more options for reducing risk such as private shelters and building to FZ, and the nationally accepted method for BAL assessments would replace the more onerous Victorian method.

I was cautiously optimistic that increased flexibility in the revised BMO would enable us, and many others, to build our homes.

THE BMO AND REGULATORY FAILURE

During the 2.5-year community campaign for changes to the BMO, the failures of the policy development process emerged. The process was authoritarian, and little regard was apparently given to independent expert advice or impacted communities. A lack of impact analysis and evidence-based reasoning resulted in regulatory failure.

EXPERT ADVICE

The VBRC had been heavily critical of state government, local government, and government agencies for a variety of shortcomings, including land use planning policies. This chapter argues that in the often irrational, defensive, and emotional period after a disaster, it is particularly important that independent experts have a major role in policy development to ensure policy reform is evidence and risk based.

The BMO was developed to replace the Wildfire Management Overlay (WMO), which had been developed in response to the 1983 Ash Wednesday bushfires. The BMO established a new BAL site assessment method and mandatory defendable space distances, which were more conservative than the nationally accepted building standards method.

The DTPLI and CFA consulted with experts during the formation of the BMO policy in 2010 and 2011. Experts advised that the proposed BMO assumptions and inputs would generate higher risk assessments, higher compliance costs, and reduced development compared to the existing national construction standards BAL assessment methodology (Bennetts, 2011; Shaw, 2011).

> *(The proposed BMO inputs)...would result in a total set-back of more than twice the value that would be required by simply applying the building standard. If this is what is proposed, it is not considered to have any basis at all*
>
> **Bennetts (2011)**

And

> *The (BMO) Working Group should avoid taking the approach of safety independent of accepted cost/benefit analysis which would lead to a legislative or policy response that exceeds the logical analysis of the risk in a true cost versus benefit sense and does not meet community and Government expectations*
>
> **Shaw (2011)**

However, in its briefing to the Planning Minister, DTPLI stated that the impact of the BMO would be an *increased* number of building developments, *lower building costs*, and *fewer* BAL FZ assessments:

> *The single site assessment process in most instances will reduce the BAL required for a particular site and the number of sites assessed as BAL FZ due to different modelling assumptions from the current WMO process. This reduction in the BAL will need to be balanced by more stringent vegetation management prescriptions to achieve an acceptable level of safety.*

> *This means that sites previously considered by the CFA as too high a risk for development may now be able to be developed, allowing landowners to build homes. Building costs for landowners may also be lower as houses are likely to be able to be built to a lower BAL.*
>
> **DTPLI (2011)**

Inconceivably, the DTPLI Minister's brief was contrary to expert opinion provided earlier in the year and provided the Minister with a false prediction of the policy impacts on landowners.

IMPACT ANALYSIS

The inputs in the national construction standards BAL assessment were scrutinized as part of the Australian Building Codes Board *Construction in Bushfire-Prone Areas—Regulatory Impact Statement*, a rigorous cost/benefit analysis, alternative policy evaluation, and stakeholder consultation process (ABCB, 2009). The VBRC also considered the national construction standards BAL inputs and "despite great scrutiny by the VBRC and many witness statements in relation to the Standard…the VBRC made no comment on the inadequacy of the (BAL inputs) FDI 100 or the Flame Temperature 1090K" (Shaw, 2011) and increasing those inputs "appears to be based on CFA's existing policy rather than scientific evidence" (Shaw, 2011).

I argue that the DTPLI and CFA's decision to alter those inputs without evidence-based reasoning, rigorous impact analysis, and cost/benefit analysis demonstrated a failure of due process. Mapping the new BAL ratings would have identified the scale of land sterilization caused by the BMO. For example, BAL mapping was conducted for the township of Wye River (Colac Otway Shire, 2014) in 2014. Over 90% of existing properties were rated BAL FZ (Fig. 14.9).

The BMO prevented BAL FZ development, therefore 95% of the township became undevelopable. The impact of a BAL FZ rating was that vacant land was sterilized and existing homes could not be upgraded or rebuilt, resulting in disastrous economic, social, and bushfire resilience outcomes. If this mapping exercise was carried out in 2011, it would have informed government of the degree of land sterilization caused by a more conservative BAL assessment method.

PRIORITY OF LIFE

The VBRC concluded that life safety must be prioritized over property protection, which became the basis of the new public safety campaign "leave early and live." The BMO policy did not allow risk to life to be addressed with the use of approved private and community shelters, mandatory evacuation, or any other measure. The development of single dwellings in high-risk areas could simply be refused, despite the growing number of technical solutions available to prioritize life safety.

Vulnerable use buildings were not included in state planning or building policy or in the National Construction Code. Existing homes were not addressed by any policy at all. Both categories of buildings presented greater risk to life than new bushfire compliant dwellings (Fig. 14.10).

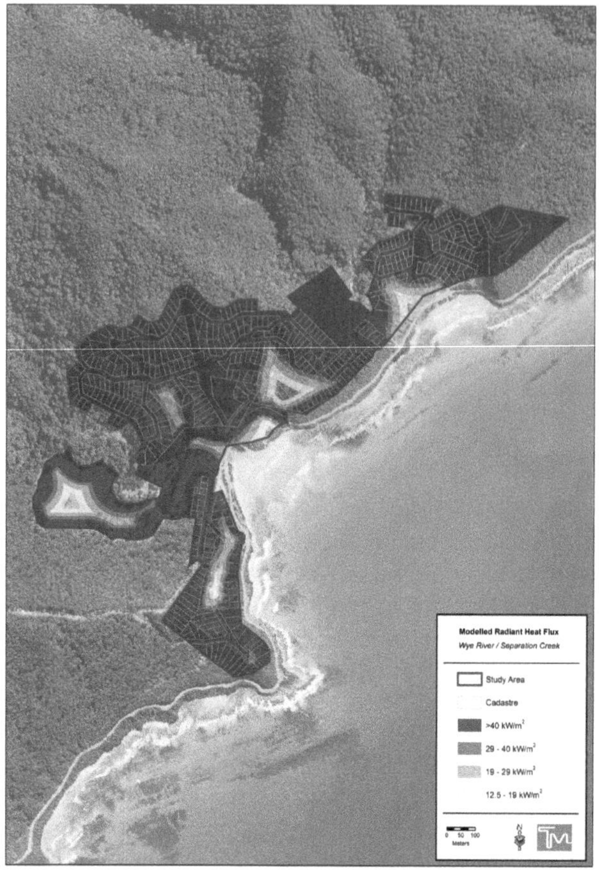

FIGURE 14.9

Wye River BMO BAL mapping.

Courtesy of Terramatrix (2013, p. 188).

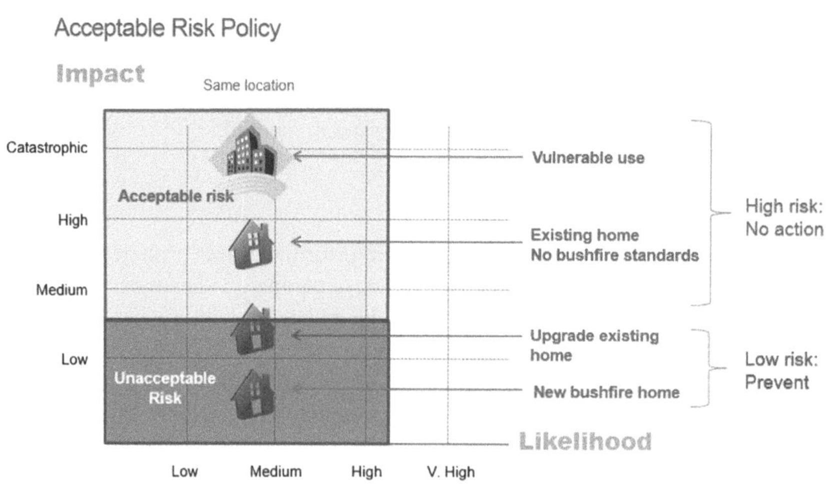

FIGURE 14.10

Acceptable risk and regulation.

From Cotter (2012).

As blocks of land became sterilized under the BMO, neighboring residents declared that they would have to stay and defend their existing homes, because the BMO would prevent them rebuilding after a bushfire. The regulation was changing human behavior toward a higher risk outcome.

Home owners wrote to the Minister to advise him of this policy outcome:

By making rebuilding very difficult or impossible these restrictions will force residents to revaluate their leave early policy and reluctantly stay to defend their home in a bushfire which could result in further lives lost.

Irving (2012)

For example, Ed Kennedy's planning permit application was refused, where he proposed to build to the highest allowable bushfire construction requirements. After his application was refused by council, Mr. Kennedy took the case to the Victorian Civil and Administrative Tribunal, where the tribunal upheld the refusal of a permit, and stated:

This (location and risk) is not a position I would want to put any family in

VCAT (2012)

Rather than being permitted to build to strict new bushfire standards, the Kennedy family now rent a dilapidated, 80-year-old weatherboard home next door to their sterilized land. The BMO policy forced the Kennedys' into a higher risk situation (Fig. 14.11).

IMPACT ON LANDOWNERS

The response from government and agencies was consistent, such as this example from DTPLI to Mr. Ben Adamson:

In the aftermath of this disaster the VBRC recommended changes to the Planning and Building controls. These have been implemented and there are circumstances, like yours where it is not responsible to support new dwellings in areas where the risk is too great

Monk (2013)

Mr. Adamson had paid AUD $272,000 for his land, which was zoned to allow the development of a dwelling, but was rendered undevelopable by the BMO. The land was unsalable as buyers would not purchase sterilized land. The site is surrounded by existing, older houses, which were allowed to be occupied, despite older homes posing a higher risk to life than newly built bushfire-rated homes. Mr. Adamson had a mortgage on the now valueless property, placing him in a negative equity situation. He was forced to rent an older home, not built to any bushfire standards in a nearby high-risk area. He had previously managed the vegetation on the site but was unable to afford to continue that maintenance, which resulted in higher fuel loads and increased risk to nearby properties. The financial and emotional strain impacted Mr. Adamson's health.

FIGURE 14.11

Illustration of regulatory failure.

From Cotter (2012).

IMPACT ON COMMUNITIES

For all townships, including those recovering from disasters such as Black Saturday, community-wide resilience and viability relies on new development to provide jobs and accommodation for locals and tourists and prevent inflated rents (RIA, 2013). Existing lawfully created subdivisions are expected to be developed over time, and that contribution to the economy is factored into local government revenue forecasts. Communities voiced their objection to the BMO being used by government to block new development and strip property rights from its citizens, which was viewed as a threat to the viability of regional areas of Victoria (Meehan, 2013).

LESSONS AND REFLECTIONS

FORMALIZING COMMUNITY ADVOCACY

A major lesson I learned was that the process of bushfire planning policy development was deeply flawed, undemocratic, inconsistent, inexpert, and lacking in strategy. Communities were not involved in the policy development that drastically impacted their viability and resilience. Community engagement and impact analysis could have prevented the adverse political, economic, and social consequences. A formal role for community advocacy in state planning policy could strengthen the democratic process and identify impacts prior to implementation. In response to this lesson, I established a national, not-for-profit organization called the Bushfire Building Council of Australia, which has a board of independent bushfire experts, and property owners can become members, free of charge. Our objective is to provide a platform for independent experts and landowners to work together to promote evidence-based policy and innovative building solutions.

EMPOWERING HIGH-RISK COMMUNITIES

Landowners and communities felt disempowered by the BMO policy and its severe impacts. For many areas where the BMO sterilized land, it may have been possible to mitigate the bushfire risk to an acceptable level for the whole community through strategic fuel management, shelters, upgrading of existing housing stock, and evacuation planning. The economic case for alternative solutions to risk management is compelling. Where building a home to meet BAL FZ requirements is estimated to add AUD $100,000 to the cost of construction, both the Victorian Community Fire Refuges and private bushfire shelters cost approximately AUD $2500 per person (MyEM, 2015; Wildfire Safety Bunkers, 2015), which may form part of a family's, or whole community's, integrated emergency management plan.

If there are circumstances where risk to life cannot be reduced to an "acceptable level," then those property owners and communities should be engaged in the decision-making process that determines their future. There are many alternatives to uncompensated land sterilization, as identified by the VBRC, to reduce development in high-risk areas (McDonald, Macintosh, & Foerster, 2013). A critical pathway to empowering communities is to inform them of their risk, both at the community level and individual property level. Statewide risk and hazard mapping is not available to the public. The mapping that is available simply shows whether a property is in a bushfire-prone area, which applies to over 90% of the state. Prospective property purchasers, existing home owners, renters, and tourists have little or no access to meaningful bushfire risk and property compliance information.

POSTDISASTER POLICY DEVELOPMENT

The postdisaster political environment is challenging; public inquiries and outraged communities often attribute blame to governments and responsible agencies. To protect themselves from future culpability or to appease public discontent, governments and agencies may implement policy reform rashly, without the normal rationale of an evidence- and risk-based approach. This chapter contends that regulatory failure, such as that caused by the BMO, can be avoided by formalizing the roles of independent experts and community advocates to ensure that policy development is balanced, rational, and follows due process, including carefully considered policy impacts.

IS THE CHALLENGE OVER?

For some lobby group members, the new BMO did not enable them to build or rebuild. Others have walked away from their land, where the financial and emotional toll was already too great. Ben Adamson has sold his land at a significant loss after the local council advised him that they would never approve development on his site, despite changes to the BMO. After 4 years of legal appeals and lobbying, Ben could not justify further time and money fighting local government (Fig. 14.12).

Lobby group members Jacqui McIntosh and Steve Exner received planning permission under the new BMO in 2014, but they are currently unable to afford to proceed as their building money was depleted due to renting for several years while fighting for changes to the BMO. Their planning permit requires high construction requirements as well as a private shelter, which is currently cost prohibitive for Jacqui and Steve (Fig. 14.13).

FIGURE 14.12

Lobby group member Ben Adamson.

From Leader Newspapers, Webb, E. (2013), Photo: Laurence Pinder/Newspix.

FIGURE 14.13

Lobby group members Jacqui McIntosh and Steve Exner.

Photo courtesy of Jacqui McIntosh.

We received a planning permit for our house at Big Hill in November 2015. We are currently still working through the building permit phase and hope to commence construction by March 2017. It has now been 5 years since we started the planning process.

The communities of Wye River and Separation Creek were devastated by bushfire on Christmas Day 2015. No lives were lost due to full emergency evacuation, but 116 properties were destroyed. Over 90% of properties in the townships had been mapped as BAL FZ prior to BMO amendments in 2014 (Terramatrix, 2013, p. 188), and rebuilding would not have been permitted under the former policy. Properties constructed under both versions of the BMO were lost in the bushfire, which highlighted inadequacies in planning policy, building standards (CSIRO, 2016), and community-wide settlement strategy. Planning policy exemptions have been established to allow rebuilding as the small lot sizes could not achieve defendable space requirements (C089, 2016). An expert panel has been established to generate alternative building solutions as existing policy rendered rebuilding unaffordable for many and would have led to fracturing of the community (WyeSep Connect, 2016). The continuing need for policy exemptions and policy "work arounds" highlights the inadequacy of the BMO to enable cost-effective risk management solutions for existing subdivisions.

REFERENCES

ABC News. (May 20, 2014). *Landholders urge Victorian Government to expedite changes to bushfire planning law.*

ABCB. (February 2009). *Final RIS–Construction in bushfire prone areas*. Australian Building Codes Board.

ABS. (November 20, 2015). *Australian national accounts: State accounts, 2014–15*. Australian Bureau of Statistics. CAT No 52220.0.

Akerman, P. (August 7, 2013). *Owners threaten action over fire plan*. The Australian.

Ainsworth, M., & Campbell, J. (May 19, 2014). *Report warns bushfire residents face financial ruin*. Herald Sun.

ANCAP. (2016). *ANCAP safety ratings explained*. Retrieved from www.ancap.com.au.

Bennetts, I. (January 30, 2011). *Review of Douglas report to CFA on criteria for planning and building in WMO areas*. Noel Arnold & Associates. Provided under FOI on 17th January 2014.

BRCIM. (2012). *Bushfires royal commission implementation monitor–Final report*.

C089. (2016). *Colac Otway planning scheme – Amendment C089: Wye River and Separation Creek Bushfire affected properties*. Gazetted 4th April 2016.

CFA. (November 2012). *Planning for bushfire Victoria – Guidelines for meeting Victoria's bushfire planning requirements*. Country Fire Authority.

Colac Otway Shire. (June 25, 2014). *Council meeting minutes*.

Cotter, K. (2012). *BMO lobby group presentation*.

CSIRO. (2016). *Wye River/Separation Creek post-bushfire building survey findings*.

DELWP. (2016). *Bushfire prone areas Victoria*. Retrieved from www.land.vic.gov.au.

Department of Justice. (2012). *Bushfire buy-back scheme. State of Victoria*. www.justice.vic.gov.au/buyback.

DTPLI. (2011). *Minister for planning brief: Single-site assessment*. Endorsed 14th July 2011. Provided under FOI on 17th January 2014.

DTPLI. (2014). *Coverage of the bushfire management overlay affected properties: Victoria*. Provided under FOI on 17th January 2014.

Fyffe, C. (September 4, 2013). *Victorian parliamentary debates, legislative assembly*.

Gunningham, S. (2015). *All the days after* (1st ed.). Sydney: Finch.

Herald Sun Editorial. (May 20, 2014). *Bushfire victims burnt too soon*. Herald Sun.

Hudson, P. (February 10, 2009). *We must tackle the evil of arson, vows Rudd*. The Sydney Morning Herald.

Irving, J. (July 16, 2012). *Letter to Shire of Yarra Ranges CEO*.

Longbottom, J. (May 20, 2013). *Landowners welcome State Government changes to controversial bushfire building laws*. ABC Online.

McDonald, J., Macintosh, A., & Foerster, A. (October 21, 2013). *Living with fire: Deciding where to build*. The Conversation.

McLeish, C. (October 29, 2013). *Victorian parliamentary debates, legislative assembly*.

Meehan. (September 3, 2013). *Up in smoke*. Upper Yarra Mail Newspaper.

Monk, J. (August 28, 2013). *DTPLI letter to Ben Adamson*.

MyEM. (2015). *BBCA innovation awards entry, blackwood community fire refuge and fire station*.

PWC. (May 2012). *Bushfire protection for vulnerable use buildings – Regulatory impact statement*. Prepared for Department of Planning and Community Development.

RIA. (August, 2013). *From disaster to renewal: the centrality of business recovery to community resilience*. Regional Institute of Australia.

Shaw, J. (2011). *Comment on UTS: CLG paper on defendable space and BAL levels for planning and building in WMO areas*. Building Commission. Provided under FOI on 17th January 2014.

Terramatrix. (2013). *Colac Otway Shire Council Bushfire Planning Policy Project – Bushfire Technical Report*.

VBRC. (2010a). Final Report, Vol. 2, *Chapter 6, Planning & building*.

VBRC. (2010b). *Final report recommendations*.

VBRC. (2010c). *Final report – Summary*.

VC57. (2009). *Victorian planning provisions 52.39: 2009 bushfire – Replacement buildings*. Gazetted 14 May 2009.

VCAT. (November 2, 2012). *Kennedy v. Cardinia SC & Ors. VCAT* 1676.

Webb, E. (May 27, 2013). *Upwey man Ben Adamson has planning application to build in Ferny Creek rejected by VCAT and Yarra Ranges Council, site deemed a fire risk*. Free Press Leader.

Wildfire Safety Bunkers. (2015). *Wildfire safety bunkers pricing schedule*. Retrieved from www.wildfiresafetybunkers.com.au/pricing.

WyeSep Connect. (2016).

FURTHER READING

BTSC. (July 19, 2011). *Bushfires taskforce steering committee*. Meeting No. 3 Minutes. Provided under FOI on 17th January 2014.

Burke, A., & Leonard, J. (2010). *Comparison and discussion of WMO and PBP*. CSIRO. Provided under FOI on 17th January 2014.

Cotter, K. (2013). *BMO and CFA referral system: Analysis and proposed solutions*.

Ginivan, J. (June 12, 2013). *Inquiry into the regulatory impact statement process* Standing committee on environment and planning legislation committee.

Paul, M. (May 28, 2014). *Victoria bushfire planning law changes welcomed by landowners*. ABC Online.

Phillips, J. (May 2, 2014). *DTPLI letter to Kate Cotter*.

Standards Australia. (2010). *AS3959 working group discussion paper – Options for construction in Flame Zone and maintenance of bushfire protection including vegetation management*. Provided under FOI on 17th January 2014.

Viney, J. (October 3, 2014). *CFA letter to Murrindindi Shire Council: Jones planning application*.

CHAPTER

Planning for Recovery: Ideas and Problematics

15

Alan March[1], Maria Kornakova[1,2]

[1]The University of Melbourne, Melbourne, VIC, Australia; [2]Massey University, Palmerston North, New Zealand

INTRODUCTION

This concluding chapter provides a summary and critical discussion of the themes addressed throughout this book. It suggests ways forward in urban planning terms by highlighting core problematics of recovery processes, while advancing overarching principles for improvement. The chapter commences with a discussion of various elements of an "ideal" settlement and its characteristics—not just in physical terms—but also in ways that join the built environment with the qualities of places that are intertwined with their constituent human and environmental systems. It discusses the challenges of recovery and relevant planning processes. The wider framework of disaster resilient settlements, as well as the theoretical and practical contributions provided in previous chapters, is drawn upon. While the scope of this book and the complexity of the topic do not allow for all elements of the framework to be covered in one single publication, we draw upon wider literature where required to illustrate some additional material missing points.

RECOVERY AND DYNAMIC SETTLEMENTS

Traditional approaches to disaster recovery were oriented mainly to redevelopment, particularly in terms of physical structures and systems, and specifically sought to find ways to improve the resistance of communities to expected possible future shocks, if the community had capacity to achieve this. While this must remain important to the resilience of communities, simply bouncing back to a predisaster state as soon as possible is understood now to be insufficient, even if new features that might improve physical resistance to features such as flood walls, or wildfire asset protections zones, have been upgraded. It has been argued and demonstrated in various studies that to simply "bounce back" often results in settlements where future disaster events might result in even more severe impacts if the wider adaptive capacities of communities are not developed as a key part of recovery (UNISDR, 2015a). With increasing numbers of disasters internationally and the increased impacts they bring to communities, it is imperative to recover in a manner that minimizes future impacts. Recovery, as a

Urban Planning for Disaster Recovery. http://dx.doi.org/10.1016/B978-0-12-804276-2.00015-3

physical development process, cannot occur independently of the social, economic, and environmental systems that support human settlements. However, it is not feasible in most cases to seek that entirely new settlements be constructed afresh. Rather, it is generally the case that recovery needs to build on the positive existing elements of settlements, after assessment of beneficial redisaster characteristics are made in parallel with modifying and improving negative characteristics. As stressed by Blakely in Chapter 5, postdisaster activities should not simply follow activities declared by politicians, but we must ensure that we are not repeating "the past when we know that the future holds new threats" (Chapter 5, pp. 63–76). He further argues that we should assist communities to recover to a state that is more or less familiar for citizens, but it should be done so intelligently to ensure that their resilience is increased toward natural or human-made disasters as well as social and economic shocks.

The building of a "new normal" that exceeds the resilience of previous settlements is often limited by the realities of recovery activities. Targeted to literally "recover" settlements focused on physical structures, these activities often do not account for other processes influencing future development and tend not to address long-term social elements contributing to resilience. These might include decreasing socioeconomic inequities, improving connectedness, and promoting social, economic, and governance inclusivity and vitality. This is particularly evident in the context of developing countries, where physical structures are often rebuilt relatively rapidly by outside agencies, while communities are left without long-term prospects. Recovery actions tend not to reflect upon the dynamics of settlements, predicted changes in demographics, vulnerable groups, and so forth. For the development of resilient settlements, however, we must take a holistic approach that will include all aspects of community development. It should be acknowledged that settlements are dynamic rather than static entities, and recovery should not obstruct or limit wider development processes. Moreover, if more care is taken to ensure that recovery-related processes are not undertaken in isolation, but include all aspects of community development, the end result will be a more adaptive and disaster resilient settlement.

How can we ensure that recovery takes into account the range of actions necessary to appropriately manage risks into the future? We argue here that urban planning is a key discipline in the recovery process that is able to bring together a range of relevant practices and understandings, which can then be used by relevant professionals as required. The following section of this chapter provides a discussion of systems we must attend in human settlements, while the section Ways Forward to Deal With Core Challenges to Plan for Disaster Recovery argues the role of planning in their "resilient recovery" and further development.

ELEMENTS OF A RESILIENT SETTLEMENT: PHYSICAL, HUMAN, AND ENVIRONMENTAL SYSTEMS

As reiterated throughout the book, the resilience capacity of human settlements depends on a range of elements, some of which are interrelated and dependent on each other, oriented around the physical particularities of places and the ways that human systems such as governance, economics, and social relations interact. We argue that planning has the potential to contribute significantly to disaster recovery by facilitating and encouraging the capacity of these elements, improving their resilience, ensuring ongoing maintenance, and improvement of overall performance (beyond disaster resilience) of human settlements as a result. Based on cases presented in this book and the wider literature, we have themed these elements into three systems that are fundamental to human settlements: physical, human, and natural. Physical elements are those aspects dealt with by the traditional built environments disciplines (e.g., buildings, roads, spatial layout, etc.); human systems including governance of settlements, its social and economic elements (e.g., community inclusion, buy back schemes); and environmental systems, referring to the manner in which ecological and natural elements interact with the settlement (e.g., floodplains or fuels for bushfires).

Traditional recovery practices have typically sought to rebuild settlements to a predisaster state. However, introduced in 2004, the concept of build back better "calls for the 'incorporation of disaster risk reduction' measures into post-disaster recovery and rehabilitation processes and use opportunities during the recovery phase to develop capacities that reduce disaster risk in the long term" (UNISDR, 2015a, p. 2). Today, resilient recovery is recognized as imperative to sustainable development, including disaster risk reduction (DRR) practices themselves. It is defined as:

> *the concept and practice of reducing disaster risks through systematic efforts to analyze and manage the causal factors of disasters, including through reduced exposure to hazards, lessened vulnerability of people and property, wise management of land and the environment, and improved preparedness for adverse events*
>
> **UNISDR (2009).**

DRR emphases the importance of understanding and responding to dynamic change processes in physical and social aspects of settlements' development, as well as the parallel development of knowledge and evidence sets. These are directly interconnected with the concept of urban resilience, understood by Meerow, Newell, and Stults (2016, p. 39) as:

> *the ability of an urban system – and all its constituent socio-ecological and socio-technical networks across temporal and spatial scales – to maintain or rapidly return to desired functions in the face of a disturbance, to adapt to change, and to quickly transform systems that limit current or future adaptive capacity.*

From the above, we consider urban systems consisting of three characteristics physical, human, and environmental. Fig. 15.1 demonstrates the conceptual relationship between these characteristics with urban resilience achieved in the conjunction of the three.

The development of disaster resilient communities is a process of systematic monitoring and adjustment of its physical, social, and natural systems to ensure that its state is evolving with changing risks and is capable of mitigating them, as well as other goals and objectives being addressed. As noted empirically though, few communities include disaster risks in their development in the first instance,

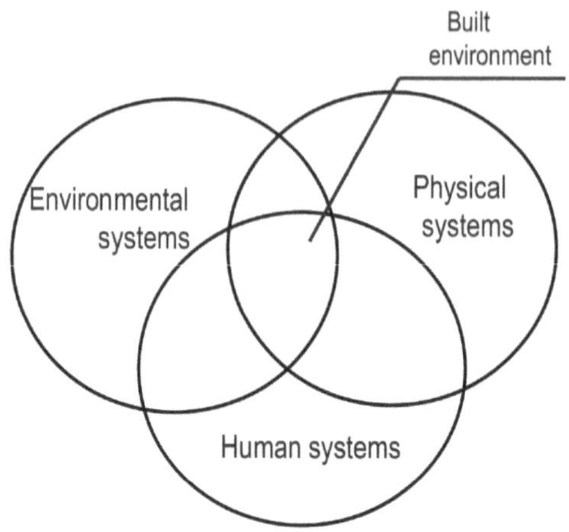

FIGURE 15.1

Conceptual representation of built environment, whereby resilience is affected by the characteristics of dynamic environmental, physical, and human systems (developed by authors).

and unfortunately, it is often only disastrous events themselves that are triggers for governments to actively integrate new relevant policies and regulations. The Swiss hazard mapping, assessment and regulation practices are an example of tragic events in 1951 acting as trigger points for strong evidence-based government intervention responding to risky land speculation and development in avalanche prone areas (see Chapter 7). This case shows the ways in which the recovery stage and associated activities can be fundamental to the development and implementation of ongoing and systematically monitored actions aiming to develop disaster resilient communities over time.

We argue that there is a need to establish a culture of prevention and improved resilience through recovery processes based on the following principles.

- Urban planning is embraced as a key aspect of good governance.
- Understanding communities as the primary object of recovery, supported by appropriate physical rebuilding.
- Urban planning is provided with a legitimate place in DRR processes, rather than being seen only as a simply regulatory mechanism—but rather is understood as a powerful tool for DRR and as a spatial process of data integration and application, beyond just regulation.
- Planning is understood as key to integrating processes and knowledge spatially, with social, economic, and ecological knowledge.
- There is a culture of initial prevention and risk reduction not just reaction and response before recovery, combined with preevent plans for recovery itself.

- Actual implementation of appropriate hazard outcomes.
- Mitigation of impacts occurring if events are unavoidable.
- The development of learning and knowledge repositories about risks and their treatments.
- Actively bringing people together and integrating subsequent actions.
- Ongoing review and monitoring of regulations, policies, and actions against previously established targets.
- Seeking to make a range of diverse community types resilient.
- Vulnerability is reduced and adaptive capability is improved.
- Good governance is understood as a legitimate outcome in itself (since it improves resilience), as the basis of good urban planning and as a starting point to integrate action.

Urban planning is particularly seen and acknowledged as a powerful tool for development of DRR and resilience. Its dynamic nature allows planning ahead and addressing principles of resilient recovery ensuring that capacity for recovery planning and monitoring is strengthened, roles of diverse stakeholders are clear, national and international policies and standards for disaster recovery strategies are developed, continuum between disaster cycle stages are maintained, and frameworks to enhance risk governance are established and promoted (UNISDR, 2015a).

Among the high-level policy assertions of the centrality of planning to DRR is the Sendai Framework for Action 2015–2030. This 15-year international agreement aims to reduce disaster risks by establishing interdisciplinary actions and acknowledges the role of urban planning in a number of key priorities (UNISDR, 2015b). More particularly, it highlights the need to promote the mainstreaming of disaster risk assessments into development and implementation of land use relevant policies; incorporating relevant hazard mapping into planning policies; aim to develop in a disaster-resistant manner where development cannot be avoided; include community in the assessment processes; promote resilience of new and existing critical infrastructure; establish and facilitate link between response, rehabilitation, and future development; and relocation of facilities and infrastructure to areas with reduced risk levels where possible. The role of land use planning is further acknowledged in the document, which highlights the importance of developing guidance for preparedness for postdisaster recovery, including lessons from other practices and exchanging knowledge and experiences. The framework addresses the dynamics of systems by promoting and reiterating the need for follow-up tools informed by changes in environment and demographic needs of communities.

It is our assertion and the one shared by many others including the International Recovery Platform (n.d., p. 64) and UNISDR (2015a) that the discipline of urban planning can deliver desirable resilience. We assert in addition, however, that urban planning is also faced with many challenges to its application in recovery. We have summarized these in six main categories set out below, most of which are impacted by the various and conflicting time scales involved in the recovery process, contrasted with the need by many government and other agencies to achieve timely results. These categories are as follows:

1. Recovery usually focuses on rebuilding, but it is actually for people
2. Equity and the harsh realities of recovery
3. Opportunities, path dependencies, and change
4. New and existing knowledge versus timely rebuilding
5. Temporal scales of "temporary" actions/or work around "Fixes" may be long lasting
6. Site specificity versus standardization and homogeneity

RECOVERY USUALLY FOCUSES ON REBUILDING, BUT IS ACTUALLY FOR PEOPLE

While recovery processes necessarily require rebuilding, cleanup, provision of basic services, and other physically oriented activities, the *purposes of rebuilding are ultimately to serve and support human needs and capabilities* within ecological and economic contexts. However, these processes occur at different temporal scales and it may be challenging to integrate human needs with the imperatives of speedy and cost-efficient recovery that is usually oriented to physical outcomes and performance indicators.

As highlighted in the cases of Sri Lanka and New Orleans, effective recovery addresses wider physical, social, and economic infrastructure matters and improves upon preexisting vulnerabilities to support long-term recovery, rather than just to provide shelter. Examples of community inclusion in processes, addressing community infrastructure needs, and supporting the local economy are discussed in Chapter 11, when local divers were provided with job opportunities during recovery to compensate for loss of their income. These cases show that key success factors include linkages being made between institutions, community, and the design and production of physical structures. The Sri Lankan case study in particular demonstrates the importance of ongoing maintenance of housing, public places (e.g., playgrounds), and infrastructure (roads) provided to the affected residents. Despite its effectiveness, this form of postoccupancy maintenance and monitoring is unfortunately rare. The Haiti example in Chapter 12 illustrates the significance of community inclusion in the rebuild stage, adding considerably to social capital while providing them with a sense of ownership and improved capacity to deal with future events.

The use of housing as a facilitator of income generation (e.g., home-based business) is often a fundamental to recovery in lower income communities, developing countries or informal settlements, as discussed in Chapters 11 and 4. This highlights the need for rebuilding to carefully attend to the needs of residents to ensure that their income is not compromised unduly. Such concerns are also applicable to wider settlements. Post-2001 recovery practices in Bhuj, India (see Chapter 8) can, in one reading of the situation, be understood as rather disruptive as it involved (voluntary) relocation of affected residents. However, the location of the new settlement was in a desirable location in terms of securing livelihood and proximity to the main urban area. Alternatively, the relocation in Banda Aceh, Indonesia, discussed in Chapter 13, can be understood as a contrasting case. Here, the village was relocated to an area identified as less tsunami prone. However, the process was carried out without careful consideration of the social and economic needs of residents, leaving them without easy access to the marina—a key generator of their livelihood. As many residents eventually returned to the original areas of their settlement that provided access to income generation activities based around fishing and other marine activities, they again developed tsunami vulnerabilities. These two cases highlight the significance of social capital and means of livelihoods for the future of a community, and the need to address or account for these early in the recovery processes, specifically when relocation is considered.

The establishment of extraordinary agencies and processes, often by "temporary" external agencies or providers, may facilitate decisive action and coordination but at the expense of building local capacity and resilience.

While the most familiar and common of such agencies are usually large NGOs, such as International Federation of Red Cross/Red Crescent Societies (IFRC) or United Nations Development Program (UNDP), the cases reported in this book reveal the significance of various other organizations that have power and potential to meaningfully support the ongoing reduction of social vulnerability. For example, contributions of the First Baptist Church organization partially supported development of housing and recreational areas in Upper Ninth Ward post-Katrina (see Chapter 11). Dutch agency Cordaid, together with a number of partners, led the Villa Rosa community-driven reconstruction project in post-2010 Haiti.

The "Make it Right" foundation in New Orleans, which still, 11 years later, continues development of housing projects in communities affected by Katrina is an example of an agency established as a response to disaster. Initiated by a wide range of individuals, it subsequently developed further and assists in the long term a wider range of vulnerable groups, such as disabled veterans in Newark, NJ, referring to New Orleans. However, it must also be acknowledged that some argue that the outputs of this foundation are not affordable for residents, suggesting they do not help the most vulnerable, and that their aesthetics do not support or reflect the community character (e.g., Alexander, 2014; Campanella & Rose, 2016; Vinnitskaya, 2013). All of these are preventing, if not reducing, building local capacity and resilience.

Recovery agencies such as Foreign Exchange Management Act (FEMA) also have significant influence upon long-term recovery and local capacities as a result of typically being responsible not only for evacuation but also temporary services, such as housing, which often become semipermanent as disasters impacts can often last longer than initially anticipated (e.g., prolonged 2001 earthquakes shocks in Ahmedabad, India) or recovery (e.g., initial cleanup) takes longer than expected due to setbacks.

The agencies and commissions established postdisaster events (e.g., Canterbury Earthquake Royal Commission post-2011 earthquakes) usually include leadership provided by high-level individuals such as external commissioners, often not expert in the field, to develop deeper understandings of reasons and suggest recommendations for system improvement. Political circumstances, time pressures, and diverse community need to add considerable pressures on these individuals and the bodies they represent. This can lead to rather hasty or "undercooked" decisions and recommendations. For example, in the post-2009 Victorian wildfire season, recommendations correctly made by the Victorian Bushfire Royal Commission to bring up-to-date building and planning codes, resulted in subsequent application of building codes that were not fully integrated and up-to-date with the latest science, and imposed unreasonable fire risk assessments across the state. The community backlash to this was considerable, since many homeowners were unable to build at all on fire-prone land, leading to a rapid political "backflip." In contrast, this mobilization of community sentiment can be seen as a positive development of local capacity and understanding of bushfire risk treatments (see Chapters 10 and 14 for more details). While it might not be available in all communities, the Sri Lanka example of Foundation of Goodness (FoG) demonstrates how locally based organizations can be utilized in a meaningful way in the recovery activities as they have knowledge and understanding of local context and needs of the community. This is evident in provision of jobs for local fishermen as a source of income in recovery period.

EQUALITY AND HARSH REALITIES

Despite our best efforts, it remains a harsh reality that *disasters are not fair* in their impacts upon different segments of the population, and the recovery process itself may not always be fair in the ways that losses, opportunities to improve after events, and opportunities for actions among individuals and organizations are reallocated during the recovery process.

Chapter 3 discusses the dependence of the recovery processes on preplanning and prior levels of equality within affected communities. If not addressed, ongoing inequality can potentially lead to differing recovery abilities and even civil unrest among residents. While not directly addressing this, Chapter 8 discusses importance of legalization of marginalized groups and addressing informality, which would not only address questions of equity but also prevent formation of new and potentially more vulnerable settlements in the future. Such transformation of systems vulnerable to future events adds to resilience of settlements, as defined by Meerow et al. (2016).

The practices of humanitarian agencies discussed in Chapter 4 demonstrate the importance of addressing challenging and complex questions of human rights for housing, which are complicated by tenure status, land ownership, cultural contexts, but most important are recovery finances and their distribution. Recovery agencies and donors tend to provide financial assistance according to assessment of damage incurred to individual properties. While this is conceptually logical, in practice it raises a series of complications, as resources are typically limited and insufficient to assist all affected. This raises questions as to whether overall and ongoing community benefits will result from distributing recovery finances equally among all affected to use as they see fit, base it upon house damage, or vulnerability of residents (Häberli, 2013).

There are no easy answers to these questions, and we do not aim to address them here in favor of careful place-based assessments being made but rather seek to acknowledge and highlight the need to address social systems beyond just recovery.

Another example of financial incentives and assistance are incidental cash flows for jobs that residents are capable of undertaking as part of recovery processes. The cobenefits here are that activities such as cleanup of debris from sites can be achieved by able residents, such as those in the post-Haiti earthquake recovery (see Chapter 4). The challenges of such exercises include limited money or work suitable for all, limitations associated with age, care responsibilities, and physical abilities to undertake a job. Furthermore, cash flow and the role of local markets have twofold elements. From one perspective, cash flow is important in the developing world as it provides an access to income for some actors, but on the other hand they can pose threats associated with hyperinflation, uneven distribution, and impacts on the value of savings.

The equity and fairness of financial aid is often challenged by land tenure, or more accurately, a lack of tenure proof or rights. While it is often considered a problem of informal settlements in the developing world, questions of tenure in relation to financial distribution are also prominent in the developed world. For example, Hurricane Katrina revealed complex and sometimes unregulated land inheritance practices in New Orleans, resulting in the inability of residents to prove their tenure status (Baab, 2008). Reflecting on this, we argue that there is a need to address fundamental planning dynamics and political questions of land ownership and to honor the local particularities of an area.

Insurance is traditionally seen as a relatively fair mean of financial assistance after a disaster event—an affected household receives compensation based on damage and premium. Such measures, however, often do not allow for improvements to be made to achieve more resilient structures and, as a result, may contribute to the overall resilience of a community. The German practices reported in Chapter 9 demonstrate State and Federal governments' dedication to recovery through provision of 80% of reconstruction costs for homeowners without insurance. While this exemplar is fairer compared to the US practice of spending most funding on infrastructure and to provide personal loans for households, there is still a potential for those with insurance to be disadvantaged, as discussed above. Thus, we consider governance of insurance policies applied in the New Zealand as fairer and inclusive as the government established mechanisms to link with insurance companies, acknowledging that it is common for poorer people to be uninsured, often due to risk profiling meaning that riskier properties incur significantly higher premiums.

OPPORTUNITIES, PATH DEPENDENCIES, AND CHANGE

The *prior quality and arrangements of governance and community interconnectedness* have significant impacts on the ability to improve resilience in recovery processes. It is tempting for decision makers and politicians during recovery to simply seek to *rebuild without building upon strengths and making fundamental changes* and improvements where needed, due to limited resources, even when recovery may offer possibilities to change approaches. In parallel however, *many settlements are in highly dynamic states* in terms of growth, change, expansion or contraction of populations, economies, impacts on the environment, and quality and type of the built form, and it may not be advisable or even possible to seek to recreate prior arrangements.

Moreover, it is not always rational to bluntly restrict or prevent development in areas with high-risk profiles. Rather, there is a need to treat them and seek solutions for reducing risks by increasing resilience. For example, Hallegatte (2011) discusses the comparative investment advantages of hazardous areas such as ports—despite being subject to floods, storm surge, waste, dangerous goods, and pollution risks—they are often key foundations of social and economic capitals of local and national importance. Thus, development should be permitted and risk reduction practices should not compromise other needs of the community, particularly if significant prior investment has been made in these areas. However, such areas tend to attract ongoing development, including housing projects, the owners and residents of which are not necessarily aware of risks associated with the location. This suggests that there is a need to ensure that governance processes include ongoing community education. However, a fundamental question remains whether development of risky but community benefit providing areas should be permitted at all. If it is permitted, then what is the minimum risk benchmark, and if not—who is responsible for existing settlements and threats to them? We cannot provide a blanket response to this paradox but rather illustrate this point as a provocation for built environment and relevant professionals to consider as part of dynamic system change management.

Prior long-term investments in settlements, such as infrastructure and services, are significant, and it is prudent and usually necessary to maximize these to the greatest extent if appropriate. While not addressed directly in this book, the importance of services such as sewerage, water, electricity, and various hazard solutions (e.g., levees) are mentioned in the majority of cases in this book. The failure of some of these can add significantly to or even cause disasters. For example, the failure of high voltage power lines was identified as a significant cause of fire ignition on February 7, 2009, in Victoria, Australia (Victorian Bushfires Royal Commission, 2010). Another example of need to invest in infrastructure was noted in the Hurricane Katrina case, when damaged infrastructure released the toxic sewage and pollutants, and residents were exposed to what has been described as a "toxic bath," leading to ongoing health problems and adding to fatalities (American Society of Civil Engineers Hurricane Katrina External Review Panel, 2007; US Army Corps of Engineers, 2009).

Transportation and in particular roads are a key infrastructure element requiring fundamental investment. Roads' spatial arrangement, network, and connectivity to other areas and between sites, capacity, quality, width, and so forth influence evacuation and response, as well as recovery and further prevention. The problematics associated with these are the limits to risk treatments imposed by landscape and natural factors, such as steep slopes in wildfire-prone areas (Kornakova & March, 2017), low-lying terrain in areas susceptible to flooding, such as parts of New Orleans (Comfort, 2006), or roads limited in width and accessibility in mountain terrains of Nepal and India. Focused on questions of social vulnerability, the recovery process after the 2013 German floods demonstrated the severity of oil contamination in affected areas and highlighting the need to address core services as part of recovery.

Developing new understandings and support may be difficult in circumstances where the capability to plan is bounded by formal decision processes that have limited scope for modification. The scope for new planning possibilities allowed by current regulatory structures may challenge the ability to modify settlements' physical and functional form. Furthermore, mass and local media, politics, and bureaucracies may confuse "truths" and possibilities in recovery. However, the Swiss case (see Chapter 7) demonstrates positive examples of political interventions into ongoing development and regulatory structures. Addressing large-scale land speculation that was facilitating risky development in avalanche-prone areas, political conviction provided a base for the enforcement of new regulatory arrangement for land management in the country and led to the establishment of land use and planning regulations, specifically targeted to address natural hazards risks.

It is common for *existing planning approaches to be maintained*, even when many of the risk profiles associated with the disaster were to some extent brought about by the planning system itself. Because planning systems are complex and cross over into many other aspects of urban management, service provision, and tenure systems, it is usually challenging to change fundamental planning approaches. Accordingly, it is common to change one or two main aspects of the planning system or to use one type of approach at the expense of an overall change. For example, it is common to change key regulations after the event, when a wider view may suggest that this would be only one aspect of a suite of planning approaches that might be used to improve resilience. This is illustrated in Victorian wildfire planning, which responded to 2009 season by updating hazard mapping systems and tightening planning regulations.

It should be noted although that *new rules and approaches, however sensible, may be unsuccessful*. The development of new approaches, understandings and regulations among experts, decisions makers, and formal bodies does not guarantee compliance and acceptance among the community. This is illustrated to some extent in Victorian planning for wildfire, where a lack of meaningful communication and consultation with residents led to a backlash against changes in regulations. In summary, the introduction of overly restrictive regulations was so unpopular that the government was concerned for its reelection chances and rapidly relaxed the controls accordingly. Other successful community engagement in a number of recovery cases from the developing world, however, demonstrates the value of successful participation exercises. For example, the relocation of sites in India (see Chapter 8) or engaging with local communities in Haiti and Sri Lanka (see Chapters 12 and 11, respectively) show how engagement builds community trust and capacity. Building techniques negotiated in Sri Lanka post-2004 tsunami demonstrate effective collaboration between government officials and local organizations resulted in alternative livelihood sources for local divers, leading to building resilience and capacities of the community.

NEW AND EXISTING KNOWLEDGE VERSUS TIMELY REBUILDING

Disasters present opportunities to deliver new and improved settlements, based on the learning that can come about by studying events and consolidating other up-to-date evidence, international scientific knowledge, and community insights. However, it is typically a time-consuming and complex task to modify underlying rules, regulations, and mechanisms for recovery and ongoing settlement management processes. For example, limited in time and under pressure by various stakeholders, decision makers in the Victorian wildfire case, documented in Chapter 10, adopted new building codes that were imperfect resulting in ongoing construction of housing stock, which is sound structurally but not adequately resistant to potential fire disasters. In contrast, the history of land use and hazard mapping of Switzerland discussed in Chapter 7 demonstrates processes required to establish new systems in a meaningful way that is also accepted by the community. While such prolonged processes might not be desirable in all cases, it demonstrates that there are benefits of extending the temporal scale of processes compared to hasty or ad hoc rebuilding.

The development and use of new and existing evidence is integral to improved resilience in recovery processes. For example, considering the ways that data were collected and used in the Swiss case mentioned above, we argue that good planning process or good governance of the disaster recovery directly depends on knowledge and application of science to manage and improve the ways settlements change over time. The careful assessment of avalanche risks and establishment of treatments ranging from restricting development, imposing design standards, and the use of avalanche barriers to protect towns and infrastructure have proven highly effective. Contrary to this, the Victorian chapter demonstrated how a relative lack of collaboration between science and decision makers could lead to the best practice being ignored in favor of "workarounds."

TEMPORARY OR WORKAROUND SOLUTIONS MAY BE LONG LASTING

Temporary recovery arrangements such as relocation may result in eventual permanence or have significant influence on the expectations of citizens for ongoing care by authorities and may disrupt economic and social resilience and connectivity. Furthermore, temporary relocation may influence choices for future locations of homes or become permanent. Initial investment costs may have been so great that after initial finances have been expended by governments and the modifications to people's lives, and the investments have made around "temporary" or extraordinary arrangements that they become permanent.

One of the documented cases of temporary housing becoming permanent is the post-Hurricane Mitch recovery in Honduras, when 6 years after the event, hundreds of affected residents remained in temporary or transitional housing (Arnold, 2006, p. 262). While life span of transitional housing is 3–5 years according to the Department for International Development (2011), complicated governance in place, limited financial aid, poor quality of construction, lack of well-established infrastructure such as sewage system, and lack or uncertainty of source of livelihood leads to health impacts, welfare dependency, difficulties with maladaptive coping mechanisms including drug and alcohol abuse, and education of children (Boano, 2013). For example, post-2010 Haiti earthquake, number of housing programs were recalled due to difficulties associated with distributing finances and working within settings of local governance, and, by 2013, more than 80,000 of affected residents were still living in transitional tents (Sanderson & Burnell, 2013).

Transitional housing is often used to provide immediate relief and shelter for affected residents. Cases such as the displacement of more than 60,000 residents in Aceh post-2004 Indian Ocean Tsunami to poorly constructed shelters for over a year after disaster (Da Silva, 2010) demonstrate the need to establish recovery planning strategies before disaster striking or immediately to work with the community on rebuilding. Supporting this point is the housing recovery post-2004 Indian tsunami in Lam Guron village, Indonesia. Compared to those villagers relocated into temporary shelters, community residents returned to their original settlement and initiated the recovery processes. Temporary housing constructed by residents was further augmented by relevant agencies through in situ upgrades and community development programs. The presence of the community at the time of rebuilding is considered as one of the key success factors for this community in comparison to other communities (Mantel, 2013).

There may also be impacts associated with temporary regulations, advice and financial aid mechanisms that establish dependency and expectations that may not be able to be maintained for long periods. For example, in the Victorian 2009 wildfire case time-limited dispensations and financial aid associated with politically motivated promises to facilitate speedy recovery created expectations that all victims would be able to rebuild with the financial assistance, whereas time limits to regulatory dispensations to rebuilding, necessary to ensure that new building in the future would achieve higher standards of fire resistance, were eventually imposed, preventing residents who psychologically took long periods to recover from rebuilding.

However, when intelligently applied temporary structures can provide alternatives for residents, such as the FEMA trailers as described in the case of post-Katrina New Orleans, in Chapter 11. While this might not be an ideal solution in a long run, it can certainly be used as a transitioning housing solution. Reconstruction of the shopping mall in Christchurch post-2011 Canterbury earthquake is an example of the successful use of temporary solutions for the economic benefit of the community (Re:START MALL, 2014). In fact, the use of shipping containers, such as in Christchurch, is increasingly understood as being attractive for transitional housing as they are relatively cheap and easy to use and adapt. While the challenges of such measures relate to both the nature of structures and the potential restriction of timely community recovery, we believe that there is a benefit of their use in temporary housing and potential further recycling as community facilities, such as those in Christchurch. The use of such containers for more than 10 years post-2003 earthquake in Bam, Chapter 8, supports this argument. On the flip side, an example of the use of adaptable core structures as described in Chapter 13 provides a positive example of targeting limited finances to be spent on a feature that can be permanent and around which adaptive change led by residents themselves can occur over time, rather than building a full temporary structure.

SITE SPECIFICITY VERSUS STANDARDIZATION AND HOMOGENEITY

Many aspects of settlement development and management over time are based on standardized and homogenous systems that, in nondisaster times, may offer certainty, efficiency, and fairness in bureaucratic or mass construction terms. However, many aspects of human settlements may actually require highly site-specific responses in terms of human and community needs and indeed to respond effectively to the risk profiles of places.

Urban planning, with its potential to deal with spatially specific matters, can align risk assessments and treatments directly with the locations that require them. This site specificity is well illustrated in the Swiss practice of applying hazard mapping to land use and zoning, thus identifying development types allowed based on risk levels and site-specific assessments. This can be understood as having relatively standardized overarching principles in place to ensure that lower tier site-specific assessments and tailored responses occur. However, it is recognized that such practices might not be possible in some areas due to the need for considerable resources to assess and treat certain risks and the need for mechanisms being in place to provide regulatory strength to enforce plans. Importantly, it is challenging to modify existing housing stock quickly, but with the use of regulatory mechanisms, it may be possible to use the potential of spatial planning to develop resilient settlements as building stock is modified over time. Wildfire planning in Victoria, described in Chapter 10, illustrates the role of planning in the modification of natural features, such as fuel loads, allowing for greater maximization of settlements, based on understanding the context of settlements at a range of scales.

While not discussed directly in this book, the case of the 2015 wildfires in Wye River and Separation Creek, Victoria, Australia, illustrates need for application of site-specific planning and design codes. Located on the scenic coastal Great Ocean Road, these two townships are located on heavily vegetated and extremely steep slopes (up to 40 degrees), have no articulated gas system and have only one road in and out. The topography of the site encourages a building typology with understories, which are often used as storage for gas vessels, timber for heating, and other often combustible objects. The current state of houses and sites that survived the fire remains risky, highlighting the need to address human understandings and response to risks, as well as their needs (Kornakova & March, 2016).

Human settlements are overall systems of economic and social subsystems that interconnect with the environment and the physical elements of settlements. Individual elements within settlements, such as structures and buildings, need to be responsive to sites' risk profiles, within the wider context and functions of a settlement and region. The adaptability of housing is one potential measure in achieving this. Adaptability can refer to structural typology, such as the mechanisms of bracing and the struts for supporting structures. The problematics of these solutions are associated with construction costs, deficient knowledge of building function and maintenance among residents, and rapid rebuild programs that often restrict meaningful innovation and change in the design of new housing stock.

Housing adaptability also refers to addressing the needs of residents and providing them with opportunities to adapt and expand as they see suitable. The core house structures described in the Indonesian case setout in Chapter 13 is an example of a structure that allows residents to expand and rebuild in manner and pace suitable for them, as well as reduces costs and time of recovery processes. This point overlaps with another category of challenge, the dynamic and interconnectedness of planning processes and systems (Fig. 15.1).

WAYS FORWARD TO DEAL WITH CORE CHALLENGES TO PLAN FOR DISASTER RECOVERY

Recovery and prevention processes overlap in settlements that undertake ongoing dynamic change—even if the "moment" and pressures of postdisaster rebuilding suggests otherwise—the actions taken in recovery set many of the fundamental risk profiles and adaptability into settlements' futures. The dynamic nature of urban planning potentially allows it to address and reflect on changes in community composition, goals, needs, and so forth over time. The inclusive and collaborative basis of democratically based urban planning allows for the development of new knowledge and its further translation to various professionals, decision makers, agencies, and the community. More importantly, as community goals, views, and desires often conflict with those of other professionals, urban planning has the means to ensure that negotiated outcomes are arrived at to satisfy all parties. Being in a position to bring together various stakeholders, agencies, and institutions, urban planning contributes to successful recovery processes by addressing "high levels of political commitment and strong institutional frameworks, which provide greater opportunity for promoting risk reduction and building resilience, as well as a greater chance for recovery and reconstruction to be implemented in an efficient and effective manner that avoids negative consequences" (UNISDR, 2015a, p. 2).

The discussion of the challenges undertaken above and as demonstrated in the cases examined in this book suggests that it is appropriate to restate the challenges addressed in the section above. They address the fundamental temporal, regulatory, and democratic aspects associated with settlements, and the need to act between individual and collective concerns. We argue that:

1. Recovery is for people and communities, even while we might focus on structures and physical outcomes.
2. Fairer and more inclusive settlements are more resilient. Facilitating improved community functions across a range of realms delivers wide benefits as well as resilience.
3. Opportunities to improve resilience across social, governance, physical, economic, and regulatory realms must be taken in predisaster planning and recovery settings.
4. The development and application of a range of knowledge types is key to resilient recovery, even if it might sometimes slow down physical rebuilding.

5. Care must be taken not to reduce long-term resilience and to unnecessarily expend resources if undertaking temporary measures.
6. Site-specific actions must be allowed if they deliver greater resilience even if standardization is encouraged by existing systems and regulation.

To address the challenges set out above, we must focus on people and ensure that the physical objects of recovery deal with ongoing social and economic needs allowing for "new normal" to be established and maintained over time. Maintenance requires not only community inclusion but also systematic "checkups" and, if necessary, interventions from professionals and agencies. We argue that these are integral to good urban planning in any case but are particularly highlighted in recovery settings. Governance of the new normal tackles questions of equity and fairness, and marginalized and vulnerable groups need to be equally included in all processes. Good governance must also address financial costs of disasters and their distribution post-disaster event. A key principle of building back better is "greater financial resilience and predictability within government to manage and respond to disaster triggered by natural hazards, and formalized strategic and resource commitments toward recovery planning, implementation and performance management" (UNISDR, 2015a, p. 3).

The devastation and losses brought by disasters, as harsh as it sounds, often open opportunities for more resilient development. We argue that by expanding spatial length of so-called window of opportunity, new paths for development can be successfully established. This can be achieved when the former or traditional processes are reconsidered and potentially changed. We must seek improved governance that will be robust and inclusive of all diverse stakeholders and their needs.

Effective recovery is timely, but we must ensure that it is based on the best knowledge available and new science, without discarding valuable traditional understandings and practices. Moreover, while the main premise of this book is to take advantage of the potential for change and improvement after disasters, we should not limit ourselves to settlements that have been affected by disasters. Rather, there is a need to continue to learn from other cases in this book and elsewhere to increase resilience and reduce vulnerability of settlements. This can be done in collaboration with other aspects and elements of the settlement processes and build upon (UNISDR, 2015a, p. 3) the principle of resilient recovery and "strengthening mechanisms for cooperation with services in areas of recovery and reconstruction that include sharing rosters of experts, capacity building, tools, bi-lateral support between countries, progress monitoring; and standardized approaches for post-disaster assessments and recovery planning frameworks."

Immediate response and recovery often includes relocation of affected residents and the provision of temporary housing solutions. Governance must ensure that these do not fall into the problem that "nothing is more permanent that the temporary" (*Greek proverb*). We should not establish ad hoc settlements that lower long-term resilience. However, if relocation is necessary and unavoidable, we must ensure that new areas provide adequate social and economic opportunities for residents, sufficient and robust core infrastructure is in place, and risk levels are minimized or managed. Finally, recovery must directly tackle and address risk profiles of the given site or an area; so standardized approaches need to be mediated with attention to understanding when site-specific responses are required.

REFERENCES

Alexander, H. (2014). *Katrina victims say Brad Pitt's charity homes are already rotting.* Chron.

American Society of Civil Engineers Hurricane Katrina External Review Panel. (2007). *The New Orleans hurricane protection system: What went wrong and why: A report.* Retrieved from Reston, VA, USA.

Arnold, M. (2006). Disaster reconstruction and risk management for poverty reduction. *Journal of International Affairs, 59*(2), 269–279.

Baab, C. H. (2008). *Emergency community block grant funds in the Gulf Coast: Uses, challenges and lessons for the future* (B546 2008c). Washington, DC: Alabama Appleseed Center for Law & Justice, Inc.

Boano, C. (2013). Post-disaster recovery planning. Introductory notes on its challenges and potentials. In A. López-Carresi, M. Fordham, B. Wisner, I. Kelman, & J. Gaillard (Eds.), *Disaster management: International lessons in risk reduction, response and recovery.* Hoboken: Taylor and Francis.

Campanella, R., & Rose, C. (July 2016). *14 to 1: Post-Katrina architecture by the numbers.* Places https://doi.org/10.22269/160712.

Comfort, L. K. (2006). Cities at risk: Hurricane Katrina and the drowning of New Orleans. *Urban Affairs Review, 41*(4), 501–516. http://dx.doi.org/10.1177/1078087405284881.

Da Silva, J. (2010). *Lessons from Aceh: Key considerations in post-disaster reconstruction.* Warwickshire, UK: Practical Action Publishing Ltd.

Department for International Development, D. (2011). *Humanitarian emergency response review.* Retrieved from.

Hallegatte, S. (2011). *How economic growth and rational decisions can make disaster losses grow faster than wealth.* Policy Research working paper, (WPS 5617). Washington, DC.

Häberli, I. (2013). Aid distribution after Hurricane Mitch and changes in social capital in two nicaraguan rural communities. In J. E. Duyne Barenstein, & E. Leemann (Eds.), *Post-disaster reconstruction and change: Communities' perspectives* (pp. 31–54). Boca Raton: CRC Press.

International Recovery Platform. (n.d.). The guidance notes on recovery: Pre-disaster recovery planning (pp. 64). Japan: International Recovery Platform Secretariat.

Kornakova, M., & March, A. (2016). The role of defendable space as a key urban design tool for bushfire risk management. *Presented at the 9th International Urban Design Conference, Canberra, Act.* Retrieved from https://protect-au.mimecast.com/s/nbpzB5S2zxQLIE?domain=urbandesignaustralia.com.au.

Kornakova, M., & March, A. (2017). Activities in defendable space areas: Reflections on the Wye river-separation creek fire. *Australian Journal of Emergency Management, 32*(1), 59–65.

Mantel, C. (2013). Ownership, control, and accountability in post-tsunami housing reconstruction processes in Aceh, Indonesia. In J. E. Duyne Barenstein, & E. Leemann (Eds.), *Post-disaster reconstruction and change: Communities' perspectives* (pp. 55–70). Boca Raton: CRC Press.

Meerow, S., Newell, J. P., & Stults, M. (2016). Defining urban resilience: A review. *Landscape and Urban Planning*, *147*, 38–49. http://dx.doi.org/10.1016/j.landurbplan.2015.11.011.

Re: START MALL (2014). About us Retrieved from http://restart.org.nz/contact-about.

Sanderson, D., & Burnell, J. (2013). Beyond shelter after disaster: Practice, process and possibilities. In D. Sanderson, & J. Burnell (Eds.), *Beyond shelter after disaster*. Hoboken: Taylor and Francis.

UNISDR. (2009). *2009 UNISDR terminology on disaster risk reduction*. Retrieved from http://www.unisdr.org/files/7817_UNISDRTerminologyEnglish.pdf.

UNISDR. (2015a). Reconstructing after disasters: Build back better. Ministerial roundtable. In *Sendai: UN World Conference on DRR* (p. 3).

UNISDR. (2015b). Sendai framework for disaster risk reduction 2015–2030. In *Paper presented at the 3rd United Nations World Conference on DRR Sendai, Japan*. http://www.unisdr.org/we/coordinate/hfa-post2015.

US Army Corps of Engineers. (2009). Performance evaluation of the New Orleans and Southeast Louisiana hurricane protection system*Executive summary and task force* (Vol. I). Retrieved from *http://biotech.law.lsu.edu/katrina/ipet/VolumeIFINAL23Jun09 mh.pdf*.

Victorian Bushfires Royal Commission. (2010). The fires and the fire-related deaths. In B. Teague, R. McLeod, & S. Pascoe (Eds.), *The 2009 Victorian Bushfires Royal Commission Final Report Victoria: Government Printer for the State of Victoria*.

Vinnitskaya, I. (2013). *The debate over making it right in the lower Ninth ward. Arch Daily*. Retrieved from http://www.archdaily.com/356483/the-debate-over-making-it-right-in-the-lower-ninth-ward.

Index

'*Note*: Page numbers followed by "f" indicate figures and "t" indicate tables.'

D

S

Kolibri 360 GmbH
Ferdinand-Jühlke-Straße 7
99095 Erfurt, Deutschland
produktsicherheit@kolibri360.de